AF564825

Evaluation and Impact Assessment

NIPA® GENX ELECTRONIC RESOURCES & SOLUTIONS P. LTD.
New Delhi-110 034

About the Authors

Dr. C. Satapathy retired as Head and Dean of Extension Education from Orissa University of Agriculture & Technology (OUAT) in the year 2002. At present he is working as Director of Amity Humanity Foundation at Bhubaneswar Odisha under the umbrella of Amity University Noida. He obtained Ph.D. Degree from Indian Institute of Agriculture Research (IARI) in the year 1981. He has teaching and research experience of 34 years. He has guided more than 30 Ph.D. Scholars and authored 14 books in various disciplines and has published more than 200 research papers in national and international journals. At present he is involved in livelihood projects.

Dr. Sasmita Panda is working as an Assistant Professor, PG Department of Management, St. Claret College (Autonomous), Bangalore. Prior to the present assignment, she worked as a Social Scientist in Amity, Bhubaneswar, under the umbrella of Amity University Noida. Besides teaching she has undertaken a number of social development projects funded by DST, State Government, Tribal Development Dept., covering livelihood systems, tribal development, education, water management, health, and hygiene. Her efficiency has been proved in designing research projects, data analysis and writing research reports. She has obtained a Ph.D. degree from Utkal University, Odisha. She has published a good number of research papers in National and International journals along with a book chapter, in addition to participating in National and International seminars and workshops.

Evaluation and Impact Assessment

Concepts and Methods

C. Satapathy
Director
Amity Humanity Foundation, Bhubaneswar, Odisha

Sasmita Panda
Assistant Professor
PG Department of Management, St. Claret College (Autonomous)
Bangalore, Karnataka

NIPA® GENX ELECTRONIC RESOURCES & SOLUTIONS P. LTD.
New Delhi-110 034

**NIPA® GENX ELECTRONIC
RESOURCES & SOLUTIONS P. LTD.**

101,103, Vikas Surya Plaza, CU Block
L.S.C.Market, Pitam Pura, New Delhi-110 034
Ph : +91 11 27341616, 27341717, 27341718
E-mail: newindiapublishingagency@gmail.com
www: www.nipabooks.com

For customer assistance, please contact
Phone: + 91-11-27 34 17 17
Fax: + 91-11-27 34 16 16

Print ISBN: 978-93-58873-06-1

ebook ISBN: 978-93-58873-19-1

Composed and Designed by NIPA®.

Preface

Monitoring and Evaluation is an important component in social research. The scholars of social science particularly are required to undergo the course to be competent in project management. We have covered all 12 units in detail and have arranged them in logical manner as per contents of each unit. We hope the book will satisfy the requirement as contents have been presented covering all aspects. The book will serve as a key book.

Authors

Contents

1

Introduction to Evaluation

Concept of Evaluation

Evaluation is the structured interpretation and giving of meaning to predict or actual impacts of proposals or results. It looks at original objectives, and at what are either predicted or what was accomplished and how it was accomplished.

Context Based Meaning and Concept

The meaning of evaluation is the way of judging or calculating the quality, importance, amount, or value of something: Evaluation of this new treatment cannot take place until all the data has been collected.

In its earliest uses (documented in the 15th century), *context* meant "the weaving together of words in language." This sense, now obsolete, developed logically from the word's source in Latin, *contexere* "to weave or join together." *Context* now most commonly refers to the environment or setting in which something (whether words or events) exists. When we say that something is contextualized, we mean that it is placed in an appropriate setting, one in which it may be properly considered.

Appropriate Time of Evaluation

Evaluation is a critically examines a program. It involves collecting and analysing information about a program's activities, characteristics, and outcomes. Its purpose is to make judgments about a program, to improve its effectiveness, and/or to inform programming decisions (Patton, 1987).

Evaluation When?

i. A problem is identified by monitoring or on-going evaluation, and it is decided to investigate further.

ii. A donor, for example a major co-founder, may request an evaluation to fit in with their own institutional requirements.

iii. An organisation decides it wants to think at a particular aspect of work, may be in relation to areas identified in its programme plan.

When, in the stage of a project, should evaluation take place?

a) Appraisal or ex-ante evaluation

b) On-going evaluation

c) Mid-Term Evaluation

d) Final Evaluation

Gathering information for monitoring and on-going evaluation should be built into project work as a continuous process. But there are special times in the project's life when more organized **'Stand Back'** evaluations take place:

a) Appraisal or ex-ante evaluation

Information collected before a project starts, or in the very early stages, helps to define what is to be done, and provides a baseline from which to measure change.

b) On-going evaluation

Monitoring indicates whether activities are being carried out as planned and what changes are happening as a result. Monitoring should be accompanied by on-going evaluation, which analyses the information in order to improve performance during a project. If a project has been conducting regular on-going evaluation, and this is well-documented, more formal evaluations may be unnecessary.

c) Mid-Term Evaluation

This may be timed to take place after a particular stage of a project, to see what has happened so far, and make adjustments for the next stage, particularly when activities are planned to go on for two years or more. This kind of evaluation tends to concentrate on management and the balance of inputs and outputs. Donors may sometimes make payments for a second-year conditional on a midterm evaluation. An annual report, written by a project manager, can be a useful evaluative document.

d) Final Evaluation

This will happen at the end of a project, to learn lessons about how the project has been implemented and the results. A final report of a project written by a project manager can be evaluative, comparing objectives with what was achieved. Ex-post evaluations. These happen sometime (often two years or more) after a project has finished. They look at impact and sustainability. They also consider broader 'policy' issues. This kind of evaluation is rare in NGOs, and more effort should be put into promoting such studies.

Program Planning

Program planning is deciding what needs to be done, and who does what, when and where. The two key elements in successful program planning are the program or project itself and the interest and involvement of group members.

Program planning is a statement of situation objective problem and solution Planning is a decision-making process.

Program Effectiveness

Effectiveness of programs relates to the level at which the activities of program produce the desired effect. It is the magnitude of targeted result achieved from the program.

How do we measure program effectiveness?

Conduct outcome evaluation by following these steps

1. Draft an Outcome Evaluation Plan.
2. Determine what information the evaluation must provide.
3. Define the data to collect.
4. Decide on data collection methods.
5. Develop and pre-test data collection instruments.
6. Collect data.

Decision Making

Decision making is the way of choosing path through collecting relevant information to arrive at alternative resolutions.

Following steps of decision-making may help to make details of ideas more thoughtful by organizing relevant data and defining alternatives. This method enhances the chances to find out the most satisfying result in term of alternatives.

Steps in Decision Making

Know well the decision to be made

First of all, we realize that we need to make a decision. Try to clearly define the nature of the decision we must do. This beginning step is very crucial and we should define clearly the kind of path is necessary to attempt evaluation.

Collect relevant data

To collect desired data, we have to take help of both internal and external sources. We can collect internal data by self-assessment and in case of external data we shall seek from outside sources may be people or documents. The data need to be pertinent, relevant and reliable.

Identify the better option

On the basis of collected information we shall have multiple paths of action. As researcher we can use our imagination to enrich the alternatives. Steps to be taken to make list of alternatives based on their merits. Additional information own imagination lead us to be right spot.

Judge the evidence

Conclusions can be drawn based on our collected information, imagination, deep thinking whether we answer to the question identified in first step. Out of multiple alternatives we can list out options based on merit. This would lead us to a satisfactory situation to arrive at conclusions. Such practices have proved effective as reported by social scientists working on different projects belonging to various economic zones. Our value system will reveal the right choice.

Select best option among alternatives

On the basis of conclusions arrived so far covering different steps we shall be in far better position to select the best one. We may combine other options to make a composite that could helpful to achieve the goal as a whole. This is skilled work that researchers or operators should learn. More the experience better would be decision.

Taking action

We have now reached at a position to translate the decision into action. We have now full information, documents to step into stage of implementation. Taking action on decision requires considerations of many things. The experience would lead the path. We are ready to as per Step 5.

Final decision

At this final stage there is need to look at decisions took so far with cause-and effect relationship. We have to judge whether we have at our hand all necessary information to move ahead. Very carefully and critically this step has to be reviewed. See that no important information is lacking to go ahead. From step 1 to step 6 needs to be reviewed.

Accountability

It is an assurance that a person or organization is evaluated on its performance or behaviour related to something for which it is responsible. The term is related to responsibility but is regarded more from the perspective of oversight. For example, an employee may be responsible for ensuring that a response to a request for proposal meets the stipulated requirements.

Types of Accountabilities

1. Corporate Accountability
2. Media Accountability
3. Government Accountability
4. Political Accountability

Corporate Accountability

Corporate accountability is about the numbers. Each publicly traded corporation must file an annual report containing audited financials. A third-party accountant reviews these reports to ensure that there are no errors or omissions.

Media Accountability

The media is compared to that in other countries because it has constitutionally guaranteed freedom from government interference. However, the First Amendment does not free the press from accountability.

When the performance of a task is substandard, there may or may not be consequences. When there's accountability, however, the employee is held responsible for successfully completing the task or explaining why they failed to do so.

Government Accountability

The role of corporate money in politics is just one of the global issues surrounding government accountability.

Political Accountability

Political accountability can relate to political contributions, how politicians spend money and political spending itself. For example, the non-partisan Centre for Responsive Politics (now Open Secrets) published an annual scorecard ranking members of Congress based on their voting records on campaign finance issues.

Impact Assessment

Impact assessment (IA) is a structured a process for considering the implications, for people and their environment, of proposed actions while there is still an opportunity to modify (or even, if appropriate, abandon) the proposals. It is applied at all levels of decision-making, from policies to specific projects.

Types of Impact Assessment

1. Global Level
2. Policy Level
3. Program and Plan Level
4. Project Level

Key types of impact assessments include global assessments (global level), policy impact assessment (policy level), strategic environmental assessment (program and plan level), and impact assessment at project level concerning to environment.

Environmental Impact Assessment

The attempt to determine impact of environment is of priority as it affects socio cultural economic and other related system of our society. Environments Impact Assessment (EIA) is under taken to know the real status of environment around us due to implementation of projects may be of any time. We evaluate it to improve the situation.

EIA is only a tool used to find out environmental, social and economic impact before deciding to start project. The Assessment also indicate possible impact in our social living system. UNEP has been emphasizing for better environment that keeps our living healthy. At early stages of the proposed project what impact will be realized is being predicted by such evaluation. New projects coming up at different countries whether consider the possible impact on the environment. It would benefit of socio-economic system for better living.

In this regard the entire world is now conscious. There may be different policies of different countries but the following steps have to be followed the fundamental consideration remain as discussed.

a) Screen out which projects require full or partial assessment and for what reasons.

b) Scope or wideness of the environmental impact that would emerge because of the new project in relation to legislative laws, international convention, knowledge of experts, involvement of public need to be determined. At this stage also there is need to discuss the alternatives or modification in project proposal so that adverse impact of the projects can be averted. This stage is very important because the project could be redesigned in length and breadth to protect environment. EIA shows such direction to make necessary modification keeping interest of people as well as project intact. Careful decision making saves the situation if this aspect is viewed sincerely.

c) Once alternatives are identified their advantages, disadvantages, possible cause and effects are to be disused to provide strength in our planning, designing and arriving at right decisions. This requires experiences as well as good scientific knowledge of the part of project authorities.

d) Preparing of report on impact with necessary details with management plan for environment. Such statement should be non-technical to convenience public about good aspects of the projects. On-Technical report may be easily understood by the public concerned.

e) It needs review again keeping public policy, international guide line, public reactions and knowledgeable experts in environmental studies. All these attempts would safeguard our valuable environment.

f) Now decision yet to made whether to proceed ahead with project or not. At this stage there is need for understanding well about public, their interest and opinion.

g) Monitor and auditing of implementation. At this final stage prediction, value of benefits, public interest, Government policies and near future impacts objects are to be critically examined and monitored.

Strategic Environmental Assessment

According to Sadler and Verheem (1996) the Strategic Environmental Assessment (SEA) is to taken as systematic, valid, rechecked and formalized assessment. Such strategy has to be accepted as it makes our environment safe creating better living condition. The Strategic Environmental Assessment (SEA) by nature covers wide range of aspects over a long time.

It can be applied at national basis, geographical basis, to stream line the focus on overall environment may be of any kind projects implemented at different levels. It is not that to stop or reduce the project but to keep eye on environmental protection at regional and national level. We need project for development but cannot overlook the importance of environment for better living. There are instances with project economy has increased but quality of life had been endangered. Our society cannot afford to pollute environment for sake of developmental projects.

SEA is commonly described as being proactive and 'sustainability driven', whilst EIA is often described as being largely reactive.

Policy Advocacy

Policy advocacy is defined as active, covert, or inadvertent support of a particular policy or class of policies. Advocacy can include a variety of activities including, lobbying, litigation, public education, and forming relationships with parties of interest.

Defining Policy Advocacy

The most basic meaning of advocacy is to represent, promote, or defend some person(s), interest, or opinion. Such a broad idea encompasses many types of activities such as rights' representation and social marketing, but the focus of this manual is on the approaches adopted by organizations and coalitions in trying to change or preserve specific government programs, that is, approaches focused on influencing decisions of public policy. In order to distinguish this from other types of advocacy activities, it is rightly called as "policy advocacy." Same is also the term we use throughout the guide to make this distinction clear.

There are many definitions of policy advocacy available from multiple authors and perspectives. At their core are a number of ideas that continually come up, characterizing policy advocacy as follows:

1. A strategy to affect policy change or action: To have a structured and sequenced plan a well thought approach is necessary to achieve the goal. Advice in this regard is to be honoured with sufficient justification.
2. A primary audience of decision makers: The final aim is to influence decision makers attempting to implement the projects. The public can be approached by the experts to work towards environmental stability and improvement. (for example, their advisors, the media, the public).
3. A deliberate process of persuasive communication: In all activities and communication tools, advocates are trying to get the target audiences to understand, be convinced, and take ownership of the ideas presented. Ultimately, they should feel the urgency to take action based on the arguments presented.
4. A process that normally requires the building of momentum and support behind the proposed policy idea or recommendation: Trying to bring a change in public policy is a relatively slow process as changing attitudes and positions requires ongoing engagement, discussion, argument, and negotiation.
5. Conducted by groups of organised citizens: Normally advocacy efforts are carried out by organizations, associations, or coalitions represent the interests or positions of certain populations, but an individual may, of course, spearhead the effort.

However, taking these basic elements outlined above a little further and emphasizing the specific challenge that we develop in this chapter, our definition is as follows:

The policy is to be implemented. For such attempt local influential bodies, network, opinion leaders and others having a say in the society can take step. The project prospers with respect to policy should accept such ideas and work on them. It will indicate that opinion of the public has valuable impact on the decision makers.

In our definition, we place a great emphasis on the idea of the transfer of ownership of core ideas and thinking. In essence, this implies preparing decision makers and opinion leaders for the next policy window or even pushing them to open one for action. The influentials do good job when decision makers accept their suggestion after being convinced about the benefits.

Objectives

The purpose of evaluation is to provide a systematic and objective assessment of a project, program, policy, or initiative to determine its effectiveness, efficiency, relevance, and sustainability.

What is the main objective of evaluation?

It involves collecting and analysing information about a program's activities, characteristics, and outcomes. Its purpose is to make judgments about a program, to improve its effectiveness, and/or to inform programming decisions (Patton, 1987).

Evaluation Types

The important types of evaluation are process, impact, outcome and summative evaluation. Before we are able to measure the effectiveness of our project, we need to determine if the project is being run as intended and if it is reaching the intended audience.

Common types of evaluations

1.	Formative Evaluation.
2.	Summative Evaluation.
3.	Process Evaluation or Implementation Evaluation.
4.	Outcome Evaluation or Effectiveness Evaluation.
5.	Impact Evaluation.
6.	Performance Evaluation.
7.	Cost-Benefit Analysis.

10. Approaches of Program Evaluation

Evaluation Criteria

1.	Relevance	Is the intervention doing the right things
2.	Coherence	How well does the intervention fit
3.	Effectiveness	Is the intervention achieving its objectives
4.	Impact	What difference does the intervention make?
5.	Sustainability	Will the benefits last?

Approaches of program evaluation

Evaluation Approaches

1.	Appreciative inquiry.
2.	Beneficiary assessment.
3.	Case study.
4.	Causal link monitoring.

5.	Collaborative outcomes reporting.
6.	Contribution analysis.
7.	Critical system heuristics.
8.	Democratic evaluation.

The following resources outline the types of evaluation approaches: outcome-based, impact, process, and participatory evaluation designs.

Extension is based upon the methods of science, & it needs constant evaluation. The positive side of the work is measured on points of the changes brought about in the knowledge, skill, and attitude & adoption behaviour of the people but not merely in terms of achievement of physical target.

Context of Principles of Extension Program Evaluation

A principle is a statement of policy to guide decisions and actions in a continued manner (Mathews). The principle is a universal truth being observed and found to be true under varying conditions and circumstances. A principle is a fundamental truth and a settled rule of action. Extension programs have the definite purpose of improving rural life through individual group and community action. Extension program planning has certain principles which hold irrespective of nature cliental and the Enterprises they may pursue.

Extension program planning should be based on facts, needs and experiences, future requirements

Working for determination, adequate information about the public and related facts to be secured. The existing situation and their situation has to be collected. The present situation is analysed and interpreted based on past experiences by taking local people into confidence. This shall help arrive at future needs.

Extension program should stand on objectives with clarity

Extension is education for action leading to development. The objectives should speak of problems of the people. It should also cover clients belonging to different situations and environment. The needs of the people should be reflected at appropriate statement for knowledge of the audience, stakeholders and all concerned. Attainable objectives are to be selected on priority based which would serve the purpose of meeting requirements.

Priority based extension approach

The problems of the rural people are many and diversified. All the problems cannot be taken up at one time. It would be difficult for extension agencies as well as funding units to attend so many problems. The time is also a limiting factor. The better option is to select felt needs on priority basis so that we can reach the destination. More over since many problems are interrelated the wise step would be to limit to problem solving on priority basis.

Resource availability and utilization approach

From practical point of view the availability of resources and their utilization techniques need to take into consideration. The program starts with resource assurance along with skills of utilization would be more fruitful at an early day. The stake holder's opinion and knowledge of the extension professional can be examined in this context.

Extension programs should have a connection at various level

The program in extension needs to be in one line like block. district, state and national level. Same philosophy is to be followed at each level. This would eliminate contradiction and conflict. Conflict free programs have more adaptability. Extension philosophy stresses for such approach.

Extension programs should involve people at the local level

Extension program is the program of people. It is by the people, for the people and with the people. At local level involvement of people is a must otherwise Extension will loose its identity.

Involvement of local institution is essential

Extension Program needs active support of local institutions and organizations. It is so because rural program needs support of related departments. Our approach should be to secure cooperation of local institution for smooth and early competition of the programs.

Definite plan of work is the core aspect of extension programs the plan of work may

Definite work plan well thought of approach and details of program are the mirror of extension program. The experience reveals that well prepared work plan leads mistake free implementation. We can never able to benefit people without such visual documents.

Evaluation and extension program

Like other programs the Extension program needs evaluation and monitoring at suitable interval. Without it we cannot state where do we stand and how to move ahead. It is also required for beneficiaries to get more support from them. More over results of evaluation draws attention of all concerned. It creates more interest among people providing scope for correction at all levels.

Equality is the core value of extension program

Extension values humanity and never allows discrimination. The benefits of program should be equally shared by program participants. It has been observed that poor people are ignored in some cases. It is against philosophy of Extension Education. All are equal in the eyes of the programs.

Role and Credibility of Evaluator, Competency and Credibility of Evaluator

Define scope and purpose

Before we start any evaluation project, we need to clarify scope and purpose. We should know goal and objectives of evaluation. What are the questions and criteria that will guide our data collection and analysis? What are the intended uses and audiences of the evaluation results? By defining our scope and purpose, we can communicate our expectations, roles, and responsibilities to clients and stakeholders. we can also avoid potential conflicts of interest, misunderstandings, or scope creep that might compromise your credibility.

Evaluations are an important aspect of education. Pre-evaluations, before educational content is delivered, help to identify where the student is currently and what level our program should be introduced. Post evaluations solidify that they understand and got the value of it.

Follow professional standards and best practices

As an educational consultant evaluator, we should adhere to the professional standards and best practices of the evaluation field. These standards and principles cover aspects such as utility, feasibility, propriety, accuracy, evaluation accountability, systematic inquiry, competence, integrity, respect, and responsibility. By following these standards and practices, one can ensure that evaluation is high-quality, ethical, and relevant.

Use appropriate methods and tools

Another way to demonstrate our credibility and expertise as an educational consultant evaluator is to use appropriate methods and tools for our evaluation design, data collection, data analysis, and reporting. We should select the methods and tools that match our evaluation purpose, questions, criteria, and context and also justify our choices and explain how they address the evaluation needs and challenges. We should use reliable, valid, and credible sources of data and evidence. We should also use rigorous, transparent, and replicable procedures for data analysis and interpretation. Finally, we should use clear, concise, and engaging formats and language for reporting our findings and recommendations.

Engage stakeholders and clients

Engaging stakeholders and clients is essential for demonstrating our credibility and expertise as an educational consultant evaluator. We should involve them throughout the evaluation process, from planning to dissemination. We should seek their input, feedback, and perspectives on the evaluation purpose, questions, criteria, methods, findings, and implications. We should also address their concerns, expectations, and interests. By engaging stakeholders and clients, you can build rapport, trust, and collaboration. We can also enhance the relevance, usefulness, and utilization of your evaluation results.

Seek Feedback and Improvement

Finally, we should seek feedback and improvement as a professional evaluator. We should solicit and welcome constructive feedback from our clients, stakeholders, peers, and mentors on evaluation work. We should also reflect on our strengths, weaknesses, challenges, and opportunities for learning and growth. We should use feedback and reflection to identify areas for improvement and action plans for enhancing our evaluation skills, knowledge, and practice. We should also seek professional development opportunities such as training, mentoring, networking, or certification. By seeking feedback and improvement, we can show our commitment, professionalism, and competence as an educational consultant evaluator.

Competency of Evaluator

Reflective Practice

Reflective practice is the ability to reflect on one's actions so as to engage in a process of continuous learning.

Technical Practice

The specific practices or techniques that are used during sprint execution to properly perform the work required to deliver features that have manageable levels of technical debt and meet the Scrum team's definition of done.

Situational Practice

Situation plays important role in selection of evaluators. Their qualification, experience, ability, outlook are taken into consideration. Because there is variation in situation the selection of evaluator varies considerably because of job requirement in particular area and location.

Management Practice

Management practices usually refers to the working methods and innovations that managers use to improve the effectiveness of work systems. Common management practices include: empowering staff, training staff, introducing schemes for improving quality, and introducing various forms of new technology.

Interpersonal Practice

The Interpersonal Practice in Integrated Health, Mental Health, and Substance Abuse Pathway focuses on total health and the connection between physical, mental and behavioural health. Integrated health care creates a comprehensive approach toward caring for people in need, resulting in both higher quality care and improved outcomes for each individual.

2

Evaluation Theories

Evaluation Theory vs Practice

Theory based evaluation is an approach to evaluation (i.e., a conceptual analytical model) and not a specific method or technique. It is a way of structuring and undertaking analysis in an evaluation. A theory of change explains how an intervention is expected to produce its results.

Theories

1.	Utilization-Focused Evaluation
2.	Values Engaged Evaluation Theory
3.	Empowerment Evaluation Theory
4.	Theory-Driven Evaluation Theory

Evaluation Theories

What are Evaluation Theories?

Evaluation theories refer to the conceptual frameworks, models, and principles that guide the systematic assessment and analysis of programs, policies, interventions, and other social phenomena. They provide a set of organizing principles and methodologies for evaluating the effectiveness, efficiency, relevance, and sustainability of various interventions and initiatives in different domains, including education, healthcare, social services, environmental protection, and public policy.

Evaluation theories draw from various disciplines, such as psychology, sociology, economics, statistics, and management, and they may emphasize different aspects of the evaluation process, such as the action of stakeholders, the criteria for success, the methods for gathering of information, examination of facts and drawing of conclusion.

The evaluation theory consists of five main components:

i) Practice,
ii) Use,
iii) Knowledge,
iv) Valuing,
v) Social programming.

Many practitioners design evaluations around methodology.

Bridging the Gap: Theory and Practices

Evaluation has emerged from a diverse array of applied social sciences. Although it is practice-oriented, there has been a proliferation of research on evaluation theory to prescribe underlying frameworks of evidence-based practice. According to Shadish, Cook, & Leviton (1991), the fundamental purpose of evaluation theory is to specify feasible practices that evaluators can use to construct knowledge about the value of social programs. This explanation of evaluation theory consists of five main components: practice, use, knowledge, valuing, and social programming.

Utilization-Focused Evaluation Theory

Michael Quinn Patton, developed Utilization-Focused Evaluation (UFE) on the premise that "evaluations should be judged by their utility and actual use" (Patton, 2013). This theoretical model should be applied when the end goal is instrumental use (i.e. discrete decision-making). UFE focuses on intended use by primary intended users. To engage primary intended users, the evaluator must identify stakeholders who have the most direct, identifiable stake in the evaluation and its results, in other words, the "personal factor" (Patton 2013). The evaluator involves intended users at every stage of the process. The ultimate purpose of UFE is programmatic improvement driven by a psychology of use.

Values Engaged Evaluation Theory

Theory is known as Values Engaged Evaluation (VEE). It emphasises value of stakeholders. They should know the value of evaluation and social wellbeing in relation to project The people and their values are of prime importance. The people should be a part of evaluation. It is important to make them understand about need and use of evaluation result. Further the theory explains as how to make people aware of value of evaluation.

Three justifications for including stakeholder values:

1. Pragmatic (i.e. increases probability of use),
2. Emancipatory (i.e. enrich stakeholders), and
3. Deliberative (i.e. all aspects of interests).

When we adopt VEE method the design and contents and benefits of the projects need to understand clearly. This is more effective than summative evaluation.

Empowerment Evaluation Theory

According to David Fetterman, self-evaluation can be used to determine empowerment. Self-appraisal and empowerment are two important aspects to conduct evaluation study. He argued that evaluation should empower stake holders to understand and accept responsibility in an organization. This can enable them for ownership of evaluation task. The job of an evaluator is a couching one. Aiming at improvement the evaluator should teach about need of evaluation and how to use evaluation tools emplacing importance of self-evaluation.

Theory-Driven Evaluation Theory

Huey Chen contributed to the theory of driven evaluation. According to him program operates in open system with attributes of input, output, outcome and effects. He was in opinion that evaluator should work with stake holders to know contents of the programs, logic, models which would be helpful in undertaking proper evaluation. He emphasized change model along with driven evaluation. To make it more clear his theory is based on change model that rests on cause effect relationship.

Conclusion

The four theoretical approaches described do not advocate a particular methodology. Evaluators can use quantitative, qualitative, or a mix of both methods for data collection and analysis. However, before considering methodology, evaluators should reflect on the theoretical frameworks that guide their practice. Although evaluation is an applied science, it is important for practitioners to be knowledgeable of theory to ensure their designs are driven by intention and purpose rather than methodological tools.

Role between Practice and Theory in Evaluation

The field of evaluation involves making judgments of quality, value and Importance to support accountability assessment, learning, and to improve performance. Traditional evaluation designs assume a high level of predictability and control. The problem is that complex programs or contexts challenge this basic assumption. Often programs deal with emergent outcomes and objectives, adaptive program processes, nonlinear theories of change and evolving stakeholder expectations. Under such complex conditions, traditional evaluation methods and tools do not allow realistic and useful representations of reality. In these instances, we need a more adaptive approach to evaluation. One that fits the environment without compromising rigor. To articulate what we have found useful in seeing patterns in complex programs, understanding the dynamics in ways that are meaningful to stakeholders and recommending.

Adaptive Actions to Improve

Impacts over time. In our work, synergy has emerged between complexity theory (through the lens of human systems dynamics) and evaluation practice (through acase study of a complex program of social change). What emerges at this generative intersection is an evaluation method that is simple, robust, rigorous and flexible enough to meet the demands of twenty-first century social change.

Evaluation theories draw from various disciplines, such as psychology, sociology, economics, statistics, and management, and they may emphasize different aspects of the evaluation process, such as the action of stakeholders, the criteria for success, the methods for data collection and analysis, and the use of evaluation results.

The Importance of Understanding Evaluation Theories

Evaluation theories provide a framework for understanding the goals and processes of evaluation, as well as the role of stakeholders in the evaluation process. Here are some reasons why understanding evaluation theories is important.

Clarifying evaluation goals: Evaluation theories can help to clarify the goals and objectives of an evaluation, and ensure that the evaluation is focused on the most important questions and outcomes.

Identifying appropriate methods: Evaluation theories can help to identify appropriate evaluation process based on the aims and destination of the evaluation.

Engaging stakeholders: Evaluation theories can help to identify and engage stakeholders in the evaluation process, and ensure that their perspectives and needs are taken into account.

Ensuring evaluation quality: Evaluation theories can help to ensure that the evaluation is conducted in a rigorous and systematic manner, and that the findings are valid and reliable.

Enhancing evaluation impact: Evaluation theories can help to ensure that evaluation findings are used to inform decision-making, improve program effectiveness, and promote social justice and equity.

Understanding evaluation theories is necessary for efficient methods of monitoring and evaluation (M&E) practice, which indicates reason of importance to study these ideas. Evaluation theories may be helpful in ensuring that assessments are focused, rigorous, and have an influential outcome because they provide a framework for understanding the objectives and procedures of evaluation.

Utilization-Focused Evaluation Theory

This theory emphasizes the importance of designing evaluations that are useful and relevant to the intended users. The focus is on identifying and addressing the information needs of stakeholders, and using evaluation findings to inform decision-making and program improvement.

Utilization-Focused Evaluation *(UFE)* is an evaluation theory developed by Michael Quinn Patton that emphasizes the importance of designing evaluations that are useful and relevant to the intended users. The focus is on identifying and addressing the information needs of stakeholders, and using evaluation findings to inform decision-making and program improvement.

UFE is based on the premise that the value of an evaluation lies in its use. Therefore, the evaluation design and methods should be tailored to the specific needs and interests of the stakeholders who will use the evaluation results. This requires a collaborative approach to evaluation, where stakeholders are involved in all stages of the evaluation process, from identifying evaluation questions to interpreting and using the evaluation results.

UFE also emphasizes the importance of building capacity for evaluation among stakeholders, so that they are able to participate in and use evaluations effectively. This includes providing training and support in evaluation methods and data analysis, as well as developing systems and processes to ensure that evaluation findings are used to inform decision-making.

UFE involves three key principles

Use-Driven: The evaluation is designed to meet the specific information needs of stakeholders and to inform decision-making.

Collaborative: Stakeholders are actively involved in all stages of the evaluation process, and their input and feedback are valued and used.

Iterative: The evaluation is viewed as an ongoing process of learning and improvement, and the evaluation design and methods are adapted as needed to ensure that the evaluation is meeting the needs of stakeholders.

UFE can be applied in a broader range of evaluation settings, including program evaluations, policy evaluations, and organizational evaluations. The goal is to ensure that the evaluation is relevant, credible, and useful to the intended users, and that it leads to positive change and improvement in the program or organization being evaluated.

Systems Theory

Systems Theory is an evaluation theory that views programs as complex systems that are made up of interdependent parts. The theory emphasizes the requirements to consider the context in which programs operate, and to know the interactions and relationships between program components and external factors.

In evaluation, Systems Theory provides a boundary for studying the relationships between program components and the broader context in which the program operates. It emphasizes the reason to know the inputs, processes, and outputs of the program, and the external factors that may influence the program's success or failure. These external factors may include economic, political, social, and cultural factors, as well as other programs or interventions that may affect the program being evaluated.

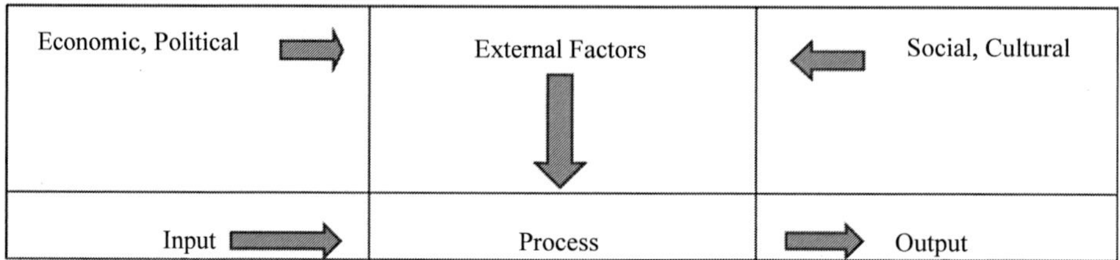

Systems Theory also emphasizes the need to consider the feedback loops and interactions between program components. This includes both the positive feedback loops that reinforce program successes, as well as the negative feedback loops that may lead to program failures. The evaluation should seek to identify and understand these feedback loops, and to use this information to inform program improvement and adaptation.

Systems Theory is useful for evaluations that are complex and multi-faceted, and that operate within a broader context. It helps to identify the interconnections between program components and external factors, and to understand how these factors may influence the success or failure of the program. Systems Theory can be used in a variety of evaluation settings, including program evaluations, policy evaluations, and organizational evaluations, and can be applied to both qualitative and quantitative data.

Empowerment Evaluation Theory

Empowerment Evaluation is an evaluation theory developed by David Fetterman that emphasizes the participation of stakeholders in the evaluation process, with the goal of promoting learning, capacity building, and empowerment. The focus is on developing the skills and knowledge of stakeholders to participate in the evaluation and to use the findings to make informed decisions.

Empowerment Evaluation involves a collaborative and participatory approach to evaluation, where stakeholders are involved in all stages of the evaluation process. This includes identifying evaluation questions, collecting and analysing data, and interpreting and using evaluation results. The goal is to build the capacity of stakeholders to participate in and use evaluations effectively.

Empowerment Evaluation involves three key principles:

1. Improvement: The evaluation is designed to promote program improvement and to build the capacity of stakeholders to participate in and use evaluations effectively.
2. Participation: Stakeholders are actively involved in all stages of the evaluation process, and their input and feedback is valued and used.
3. Social Justice: The evaluation is grounded in a social justice framework, which emphasizes the importance of promoting equity, inclusion, and empowerment.

Empowerment Evaluation is useful for evaluations that aim to promote social change and empower communities or organizations. It is often used in evaluations of community-based programs, where stakeholders have a vested interest in the program's success and are motivated to participate in the appraisal process. It is also used in evaluations of programs that serve marginalized or underrepresented populations, where the goal is to build capacity and promote equity and social justice.

Logic Model Theory

Logic Model Theory is an evaluation theory that emphasizes the importance of developing a clear and logical framework for program planning and evaluation. The theory emphasizes the need to clearly articulate the inputs, activities, outputs, outcomes, and impact of a program in a logical and coherent way, to facilitate program planning, implementation, and evaluation.

In a logic model, the program's inputs are the resources that are available to the program, including funding, staff, and other resources. The activities are the program's interventions, or the actions taken to achieve the program's goals. The outputs are the direct products or services of the program, the number of participants served or the events held. The outcomes are the short-term and intermediate-term changes that occur as a result of the program, such as changes in knowledge, attitudes, or behaviours. The impact is the long-term change that occurs as a result of the program, such as improved health outcomes or reduced rates of crime.

The logic model provides a visual representation of the program and the relationships between the program's components. It helps to clarify the program's goals and objectives, and to identify the inputs and activities that are most likely to lead to the desired outcomes and impact. The logic model can also be used to guide program implementation and to monitor and evaluate program performance.

Logic Model Theory is useful for evaluations of complex programs or initiatives, where a clear and logical framework is necessary to guide program planning and evaluation. It is often used in program evaluations, policy evaluations, and organizational evaluations, and can be applied to both qualitative and quantitative data.

Logical Framework

A logical framework, also known as a log frame, is a tool used in monitoring and evaluation (M&E) to help programs and organizations develop a systematic and structured approach to planning, implementing, and evaluating projects. A logical framework consists of a matrix that outlines the key components of a project, including the project goal, objectives, activities, indicators, and means of verification.

The logical framework approach involves a step-by-step process of developing a project plan and monitoring progress towards achieving project goals and objectives. The process typically involves four key steps:

1. Problem analysis: Identify the problem or need that the project is intended to address and the factors that contribute to the problem.
2. Objective analysis: Identify the specific objectives of the project and the activities that will be undertaken to achieve these objectives.
3. Indicator selection: Identify the indicators that will be used to measure progress towards achieving the objectives.
4. Means of verification: Identify the sources and methods of data collection that will be used to measure progress towards achieving the objectives.

In general, the use of a logical framework approach can assist programs and organisations in developing a distinct and well-structured plan for the accomplishment of their goals. Additionally, this approach can assist in ensuring that program activities and outcomes are aligned with program objectives. A logical framework enables companies to more effectively monitor and analyse the development of their initiatives, as well as to make decisions that are informed by data in order to increase the efficacy and impact of their programs.

Causal Loop Diagrams

Causal loop diagrams (CLDs) are a tool used in systems thinking and evaluation to visualize the complex causal relationships that exist between different components of a system. A causal loop diagram consists of a set of interconnected loops that represent the relationships between different components of a system, including the feedback loops that drive system behaviour.

CLDs are useful for understanding the complex interactions that exist within a system and for identifying the key drivers of system behaviour. They are often used in program evaluation to help program managers and evaluators understand the factors that contribute to program success or failure, and to identify potential areas for improvement.

Causal loop diagram typically involves several steps, including:

1. Identifying the key components of the system: This involves identifying the key elements of the system that are relevant to the program or intervention being evaluated.
2. Mapping the causal relationships between components: This involves identifying the causal relationships between different components of the system and representing these relationships in the form of interconnected loops.
3. Identifying feedback loops: This involves identifying the feedback loops that exist within the system and understanding how these feedback loops drive system behaviour.
4. Analysing the diagram: This involves analysing the diagram to identify the key drivers of system behaviour and to identify potential areas for improvement.

Overall, causal loop diagrams are a powerful technique for knowing systems and for identifying the factors that contribute to program success or failure. By visualizing the causal relationships and feedback loops that exist within a system, program managers and evaluators can better understand the drivers of system behaviour and make data-driven decisions to improve program effectiveness and impact.

Stock and flow diagrams

Stock and flow diagrams are a tool used in systems thinking and evaluation to represent the dynamic relationships that exist between different components of a system. Stock and flow diagrams are used to visualize the inflows and outflows of materials, energy, or other resources within a system over time.

A stock and flow diagram consists of two types of components: stocks and flows. Stocks represent the accumulation of resources within a system, and the amount of water in a reservoir, the people in a population, or the amount of money in a bank account. Flows represent the movement of resources within the system, and the flow of water into or out of a reservoir, the coming out of people out of a population, or the incoming of money into or out of a bank account.

Stock and flow diagrams are useful for understanding the behaviour of complex systems over time and for identifying the key drivers of system behaviour. They are often used in program evaluation to help program managers and evaluators understand the factors that contribute to program success or failure, and to identify potential areas for improvement.

The process of developing a stock and flow diagram typically involves several steps, including:

1. Identifying the key components of the system: This involves identifying the key stocks and flows within the system that are relevant to the program or intervention being evaluated.
2. Mapping the relationships between components: This involves identifying the relationships between different stocks and flows within the system and representing these relationships.
3. Analysing the diagram: This involves analysing the diagram to know the key drivers of system behaviour and to find out potential areas for improvement.

Concept Map

It is a tool used in evaluation and reason to visually represent the relationships between different concepts or ideas. These are useful for organizing and synthesizing information, and for identifying the key relationships and themes that exist between different concepts.

The map consists of nodes, which represent different concepts or ideas, and links or connectors, which represent the relationships between these concepts or ideas. The links can be directional or bidirectional and can represent different types of relationships, such as causal relationships, hierarchical relationships, or associative relationships.

The maps are good for a range of evaluation and research activities, including literature reviews, program planning and development, and data analysis. They can help to identify the key themes and relationships within a complex set of data or ideas and can provide a visual representation of these relationships to know and communicate.

Concept map typically involves several steps, including:

1. Identifying the key concepts or ideas: This involves identifying the key concepts or ideas that are relevant to the evaluation or research question being addressed.
2. Mapping the relationships between concepts: This involves identifying the links to different concepts or ideas and representing these relationships in the shape of a concept map.
3. Analysing the concept map: This involves analyzing the concept map to identify the key themes and relationships within the data or ideas being analysed.

A concept map is a useful tool for organising and synthesising information, as well as locating the core ideas and relationships included within complicated data or concepts. These are good assessors and researchers because they provide a visual representation of the links between various concepts or ideas. This helps both parties better comprehend and explain the results of their work.

Network Map

A network map is a tool used in evaluation and research to visually represent the relationships between different actors, organizations, or entities within a system. Network maps are useful for understanding the structure and dynamics of complex systems, and for identifying the key actors and relationships within these systems.

A network map consists of a set of nodes, which represent different actors or entities, and a set of links or connectors, which represent the relationships between these actors or entities. The links can be directional or bidirectional and can represent different types of relationships, such as collaboration, communication, or influence.

Network maps are good for a wide range of evaluation and research activities, including stakeholder analysis, program planning and development, and policy analysis. For the sake help to identify the key actors and relationships within a system and can provide insights into the structure and dynamics of these systems.

Network map typically involves several steps, including:

1. Identifying the key actors or entities: This involves identifying the key actors or entities within the system that are relevant to the evaluation or research question being addressed.
2. Mapping the relationships between actors: This involves identifying the connections between different actors or entities and representing these relationships in the shape of a network map.
3. Analysing the network map: This involves analysing the network map to search the key actors and relationships within the system, and to identify potential areas for improvement.

A network map is an effective method for gaining knowledge of the structure and dynamics of complex systems, as well as for locating the important players and links that are present within these systems. Network maps can assist evaluators and researchers in better understanding and communicating the findings of their work, and in identifying potential strategies for increasing the efficiency of programs or systems. This is achieved by visually representing the relationships that exist between various actors or entities.

Path Model

A path model is a statistical tool used in evaluation and research to analyse the relationships between multiple variables or factors. Path models are useful for understanding the interwoven connections between different variables and for identifying the key drivers of program or system outcomes.

A path model consists of a set of variables, which represent different factors that may influence program or system outcomes, and a set of arrows or paths, which represent the hypothesized relationships between these variables. Path models can be used to test hypotheses about the causal relationships between variables and to identify the key variables that are most strongly related to program or system outcomes.

Path models are useful for a range of evaluation and research activities, including program evaluation, policy analysis, and impact evaluation. They can help to identify the key drivers of program or system outcomes and to identify potential areas for improvement.

A path model typically involves several steps, including:

1. Identifying the key variables: This involves identifying the key variables that are relevant to the program or system being evaluated.
2. Specifying the hypothesized relationships: This involves specifying the hypothesized relationships between the variables, based on theory or prior research.
3. Estimating the path coefficients: This involves estimating the strength and direction of the relationships between variables using statistical methods.
4. Evaluating the model fit: This involves evaluating the overall fit of the path model to the data, and making any necessary modifications to improve the fit.

Overall, path models are a powerful tool for understanding the complicated links between different factors and for identifying the key drivers of program or system outcomes. By identifying the key variables that are most strongly related to program or system outcomes, path models can help evaluators and researchers to make data-driven decisions to improve program effectiveness and impact.

Realist Evaluation Theory

Realist Evaluation is an evaluation theory that focuses on understanding how interventions work in different contexts, and for whom, and why. The theory emphasizes the need to identify the underlying mechanisms that explain how and why a program or intervention works, and how these mechanisms interact with the context in which the program operates.

Realist Evaluation Theory is based on the assumption that programs are complex and are influenced by a variety of factors, including the context in which the program operates, the mechanisms that underlie the program, and the interactions between the program and its stakeholders. Realist Evaluation seeks to identify the context-mechanism-outcome configurations that explain how and why a program works or doesn't work.

Realist Evaluation involves several key steps:

1. Developing program theories: The first step in Realist Evaluation is to develop program theories that explain how the program is expected to work, and under what conditions.
2. Testing program theories: The second step is to test the program theories by collecting data on the program and analysing it to identify the context-mechanism-outcome configurations.
3. Refining program theories: The third step is to refine the program theories on the basis of results of the evaluation, and to identify areas for improvement and further testing.

Realist Evaluation is useful for evaluations of complex programs or interventions, where a clear understanding of the mechanisms that underlie the program is necessary to identify the factors that contribute to the program's success or failure. It is often used in evaluations of health interventions, social programs, and education programs. Realist Evaluation can be applied to both qualitative and quantitative data, and can be used to inform program design, implementation, and evaluation.

Nested and Hybrid Models

There are many different sorts of models, some of which have been shown in the sections that came before this one., there is potential of combining the various types of models. Each distinct kind of program theory has its own set of benefits, which, when combined with the others, may be of great use in the development and assessment of programs. In what follows, we are going to look at two different possibilities: nested models and hybrid models.

Nested and hybrid models are two types of statistical models used in evaluation and research to analyse complex data and relationships between variables. A nested model is a type of hierarchical model in which lower-level units of analysis (such as individuals) are nested within higher-level units (such as organizations or communities). Nested models are useful for analysing data with a hierarchical structure, and for accounting for the non-independence of observations within higher-level units.

A hybrid model is a type of statistical model that combines elements of different types of models, such as linear regression, logistic regression, or path analysis. Hybrid models are useful for analysing complicated connections between variables that cannot be adequately explained by a single type of model.

Nested and hybrid models are useful for a range of evaluation and research activities, including program evaluation, impact evaluation, and policy analysis. They can help to identify the key drivers of program or system outcomes and to identify potential areas for improvement.

Nested or hybrid model typically involves several steps, including:

1. Identifying the key variables: This involves identifying the key variables that are relevant to the program or system being evaluated.
2. Selecting the appropriate model: This involves selecting the appropriate statistical model based on the structure of the data and the research question being addressed.
3. Estimating the model parameters: This involves estimating the model parameters using statistical methods, such as maximum likelihood estimation.
4. Evaluating the model fit: This involves evaluating the overall fit of the model to the data, and making any necessary modifications to improve the fit.

Overall, nested and hybrid models are powerful tools for analyzing complex data and relationships between variables, and for identifying the key drivers of program or system outcomes. By identifying the key variables that are most strongly related to program or system outcomes, nested and hybrid models can help evaluators and researchers to make data-driven decisions to improve program effectiveness and impact.

Choosing the Right Evaluation Theory for Program or Intervention

It is essential to select the appropriate evaluation theory when evaluating a program or intervention in order to guarantee that the assessment will be relevant, successful, and have an impact. While choosing an assessment theory, the following are some considerations to keep in mind.

1. Program goals and objectives: The evaluation theory should align with the goals and objectives of the program or intervention. For example, utilization-focused evaluation may be appropriate for programs that prioritize the use of evaluation findings to improve program effectiveness.
2. Program context: The evaluation theory should be appropriate for the context in which the program or intervention is being implemented. For

example, developmental evaluation may be appropriate for programs that are being implemented in complex or rapidly changing environments.

3. Stakeholder needs: The evaluation theory should take into account the needs and perspectives of program stakeholders, including program participants, funders, and other key stakeholders.
4. Available resources: The evaluation theory should be feasible and practical given the available resources, including time, budget, and staff capacity.
5. Ethics and values: The evaluation theory should align with the ethical principles and values of the program or intervention, including principles of social justice and equity.

In general, picking the appropriate evaluation theory is vital for ensuring that the assessment is relevant, successful, and has an influence on the intended audience. Program managers and evaluators are able to select the evaluation theory that is most suitable for the program or intervention they are evaluating by first considering the goals and objectives of the program, the context of the program, the needs of stakeholders, the resources that are currently available, and ethical principles and values.

Using Evaluation Theories to Enhance M&E Practice

Evaluation theories can serve as a powerful tool for enhancing output of monitoring and evaluation (M&E) practice. By providing a framework for understanding the goals and processes of evaluation, evaluation theories can help program managers and evaluators to clarify evaluation goals, select appropriate evaluation methods, engage stakeholders, ensure evaluation quality, promote social justice and equity, and promote learning and improvement.

Here are some strategies for using evaluation theories to enhance M&E practice:

1. Clarify evaluation goals: Evaluation theories can help to clarify the goals and objectives of the evaluation, and ensure that the evaluation is focused on the most important questions and outcomes.
2. Select appropriate evaluation methods: Evaluation theories can help to identify appropriate evaluation methods and techniques considered on the goals and aim of the evaluation.
3. Engage stakeholders: Evaluation theories can help to identify and engage stakeholders in the evaluation process, and ensure that their perspectives and needs are taken into account.
4. Ensure evaluation quality: Evaluation theories can help to ensure that the evaluation is conducted in a rigorous and systematic manner, and that the findings are valid and reliable.

5. Promote social justice and equity: Evaluation theories can help to promote social justice and equity by ensuring that evaluation processes and findings are inclusive, responsive, and equitable.
6. Promote learning and improvement: Evaluation theories can help to promote learning and improvement by ensuring that evaluation findings are used to inform decision-making, improve program effectiveness, and enhance social impact.

Program managers and evaluators may increase the value and impact of their M&E efforts, as well as the results of programs and interventions if they integrate evaluation theories into M&E practice.

Conclusion on Evaluation Theories

As discussed, the selection of an evaluation theory depends on the program being evaluated, the stakeholders involved, and the evaluation goals and objectives.

"A program evaluation theory is a coherent set of conceptual, hypothetical, pragmatic, and ethical principles forming a general framework to guide the study and practice of program evaluation."

"Theories provide guidance in determining the purposes for evaluations, as well as in defining what we consider to be acceptable evidence for making decisions in an evaluation" Mertens and Wilson 2012:37.

Each theory has its own strengths and weaknesses but all can be used to inform decision-making. Theories differ according to their assumptions on value, use, and role of evaluation. However, differing perspectives should not be understood as exclusive of each other, rather as complementary, since all theories address elements of the evaluation practice just like a mosaic. Hence, a full understanding of their differences and similarities is useful and important to draw the whole picture of evaluation. With this in mind, it is important to recognize that any evaluation should be of the context oriented and is being conducted in order to ensure that the most effective approach is taken.

Three broad categories of theories that evaluators use in their works

Our view is that there are three basic elements in considering evaluation theories: use, methods, and valuing. All theorists are concerned with the methods that will be employed in conducting the evaluation.

The three "roots" of the evaluation theory tree-social accountability, systematic social inquiry, and epistemology-serve as a foundation for evaluation work; each has served as an important impetus for evaluation work, and therefore, each has supported the development of the field in different ways.

Program Theory

Program theories also referred to as logic models or theories of change (and other slightly different terms), are widely used in evaluation. A program theory can be broadly defined as a visual and narrative description of the main program inputs, activities, outputs, and desired outcomes.

For example, a program to reduce aggression based on social learning theory could have an underlying theory like this: "IF program mentors model strategies for avoiding alcohol, tobacco, or other drugs in social situations and provide opportunities for participants to practice these strategies.

All human service programs are designed to make a difference in the lives of people or to improve our society. This resource guide discusses program theory and logic models. Program theory explains why a program is expected to work and a logic model illustrates a program theory.

Social Science Theory

Social science theory is 'a systematic explanation for the observed facts and laws that relate to a specific aspect of life'. There is a sense in which we develop theories all the time in our everyday lives, for example, to account for someone's behaviour or explain certain problems which we face.

These three theoretical orientations are:

i) Structural Functionalism,
ii) Symbolic Interactions, and
iii) Conflict Perspective.

To understand a theoretical orientation in any profession it is critical to understand what is meant by the term theory.

What is theory?

Very simply, a *theory* is a viewpoint or perspective which is explanatory. According to Babbie (1989, p. 46), social science theory is 'a systematic explanation for the observed facts and laws that relate to a specific aspect of life'.

There is a sense in which we develop theories all the time in our everyday lives, for example, to account for someone's behaviour or explain certain problems which we face. Professionals, such as information service providers, also construct viewpoints or theories. For example, suppose staff from a particular public library observe that the children's collection of the library is used less than would be expected (or they believe that is the case). Their theory of why this is so could include the fact that there are very good school libraries in the area; and that the area is basically working class with the result that parents do not encourage their children to use the library as much as middle-class parents do. They might develop other theories about what can be done to change this situation.

Check list in Evaluation

An evaluation checklist is a list for guiding an enterprise to success (formative orientation) and/or judging its merit and worth (summative orientation). Sound checklists can have profound evaluative applications.

A checklist is a list for convenient checking and reference. An evaluation checklist is a list for guiding an enterprise to success (formative orientation) and/or judging its merit and worth (summative orientation).

Values Engaged Evaluation (VEE)

VEE seeks help poor people otherwise deprived through knowledge inputs without other considerations. This approach is considered as democratic as the evaluators have to include all stake holders.

Values-engagement in evaluation involves both describing stakeholder values and prescribing certain values. Describing stakeholder values is common practice in responsive evaluation traditions. Prescribing or advocating particular values is only explicitly part of democratic, culturally responsive, critical, and other openly ideological traditions in evaluation, but we argue that it is implicit in all evaluation approaches and practices.

Empowerment Evaluation

Empowerment evaluation (EE) is an evaluation approach designed to help communities monitor and evaluate their own performance. It is used in comprehensive community initiatives as well as small-scale settings and is designed to help groups accomplish their goals.

What is an example of empowerment evaluation?

In one example, an Empowerment Evaluation was conducted to determine the result of a program aimed at increasing women's political participation in Pakistan. The evaluation team worked with local stakeholders to collect data, analyse findings, and develop action plans to address identified challenges.

Empowerment Evaluation is a collaborative and participatory approach to evaluation that focuses on building the capacity of individuals, groups, and communities to take control of their own development and achieve their goals. This approach, developed by David Fetterman in the 1990s, is based on the belief that the people who are most affected by a program or intervention are the ones who are best able to evaluate its effectiveness and determine how to improve it.

The goal of Empowerment Evaluation is to empower individuals and communities to become active participants in the evaluation process, rather than passive recipients of external evaluation. By engaging stakeholders in the evaluation process, it seeks to build trust, increase accountability, and promote learning and improvement.

The principles of Empowerment Evaluation can be applied to a wide range of evaluation activities, including program planning, implementation, and evaluation. The approach can be used to evaluate the effectiveness of existing programs, as well as to design and implement new programs in collaboration with stakeholders.

Empowerment Evaluation is a theory and approach to evaluation that seeks to empower individuals, groups, and communities to take control of their own development and achieve their goals through collaborative and participatory evaluation processes.

Key Principles of Empowerment Evaluation

These key principles of Empowerment Evaluation guide the approach to evaluation, emphasizing the importance of collaboration, participation, and empowerment. By applying these principles, Empowerment Evaluation seeks to promote sustainable development and positive social change.

Democratic Participation: Empowerment Evaluation emphasizes the importance of democratic participation in the evaluation process. This means engaging all stakeholders in the evaluation process, including program participants, staff, and community members.

Community Ownership: Empowerment Evaluation is based on the principle that communities have the knowledge and resources necessary to solve their own problems and achieve their own goals. Therefore, the approach emphasizes the importance of community ownership and control.

Capacity Building: It seeks to build the capacity of individuals and communities to take control of their own development and achieve their goals. This means providing resources, training, and support to stakeholders to enable them to participate fully in the evaluation process.

Social Justice: Empowerment Evaluation is grounded in the principles of social justice and equity. The approach seeks to promote the empowerment of traditionally marginalized groups, such as low-income communities, women, and people of colour.

Culturally Responsive: Empowerment Evaluation recognizes the importance of culture and context in the evaluation process. The approach is culturally responsive and takes into account the diversity of stakeholders and their unique perspectives and experiences.

Learning and Improvement: Empowerment Evaluation is focused on promoting learning and improvement, rather than simply assessing outcomes. The approach emphasizes the importance of feedback and reflection in the evaluation process.

Transparency and Accountability: Empowerment Evaluation is transparent and accountable to stakeholders. The approach seeks to build trust and increase accountability by sharing information openly and involving stakeholders in decision-making processes.

Function of Evaluator and in Empowerment Evaluation

The function of the evaluator in Empowerment Evaluation is different from traditional evaluation approaches. In Empowerment Evaluation, the evaluator is not an external expert who evaluates the program or intervention from a distance. Instead, the evaluator plays an active and facilitative role in the evaluation process, working in collaboration for capacity building and promotes empowerment.

Function of evaluator in Empowerment Evaluation includes:

1. Facilitation	The evaluator facilitates the evaluation process, working with stakeholders to develop evaluation plans, design data collection methods, and interpret data.
2. Capacity Building	The evaluator builds the capacity of stakeholders to participate fully in the evaluation process. This includes providing training and support to enable stakeholders to collect and analyse data, and to use evaluation results to make informed decisions.
3. Empowerment	The evaluator promotes empowerment by facilitating stakeholder participation, providing feedback and support, and building trust and accountability in the evaluation process.
4. Contextualization	The evaluator takes into account the unique cultural, social, and political context of the program or intervention being evaluated, and works with stakeholders to ensure that the evaluation process is culturally responsive and relevant
5. Reflection	The evaluator encourages reflection and learning throughout the evaluation process, working with stakeholders to identify strengths, weaknesses, and areas for improvement. The evaluator is accountable to stakeholders, sharing information openly and involving stakeholders in decision-making processes.
6. Accountability	The evaluator is accountable to stakeholders, sharing information openly and involving stakeholders in decision-making processes.

The function of the evaluator in this case is to facilitate the evaluation process, build the capacity of stakeholders, promote empowerment, contextualize the evaluation, encourage reflection and learning, and maintain accountability to stakeholders. By playing an active and facilitative role, the evaluator supports stakeholders to take control of their own development and achieve their goals.

Empowerment Evaluation Process: A Democratic & Inclusive Approach

The Empowerment Evaluation is a democratic and inclusive approach to evaluation that seeks to strengthen stakeholders to take control of their own development and achieve their goals. The process involves a series of steps that are designed to engage stakeholders in the evaluation process, promote learning and improvement, and build capacity for sustainable development.

1. Establish a partnership
2. Identify the purpose and scope of the evaluation
3. Develop evaluation questions and methods
4. Collect data: Analyze data
5. Share findings and promote learning
6. Disseminate findings

Empowerment Evaluation in Practice: Examples from Different Contexts

Empowerment opportunity is a flexible and adaptable approach that to be applied in a variety of contexts, from community-based programs to international development initiatives. Here are some examples of Empowerment Evaluation in practice:

Community-Based Programs: Empowerment Evaluation has been widely used in community-based programs aimed at addressing social and health disparities. In a study of a community-based diabetes prevention program in an African American community, the evaluation team used Empowerment Evaluation principles to involve community members in the monitoring and evaluation process, building their capacity to collect and analyze data, and empowering them to make informed decisions about the program's future.

International Development: Empowerment Evaluation has also been used in international development initiatives, including efforts to promote gender equity and women's empowerment. In one example, an Empowerment Evaluation was conducted to determine the impact of a program aimed at increasing women's political participation. The evaluation team worked with local stakeholders to collect data, analyse findings, and develop action plans to address identified challenges.

Educational Programs: Empowerment Evaluation can also be applied in educational programs, including efforts to improve student achievement and promote equity. In one example, an Empowerment Evaluation was carried out in a middle school in a low-income community to find out the effect of a program aimed at improving student performance in math. The evaluation team worked with teachers, students, and parents to collect data, analyze findings, and develop action plans to address identified needs.

Organizational Development: Empowerment Evaluation can also be used in organizational development initiatives, including efforts to promote organizational change and development. In one example, an Empowerment Evaluation was conducted to assess the effectiveness of an organization's diversity and inclusion initiatives. The evaluation team worked with stakeholders to collect data, analyze findings, and develop action plans to address identified challenges and promote a more inclusive organizational culture.

In all of these examples, Empowerment Evaluation was used to engage stakeholders in the evaluation process, promote learning and improvement, and build capacity for sustainable development. By empowering stakeholders to take control of their own development and achieve their goals, Empowerment Evaluation supports positive social change and promotes equity and inclusion.

Advantages and Challenges of Empowerment Evaluation

Empowerment Evaluation offers many advantages, including promoting partici-pation and ownership, building capacity, promoting learning and improvement, and supporting sustainable development. However, the process also poses challenges, including the need for skilled facilitation, resource constraints, and resistance to change.

Advantages

Empowerment Evaluation is inclusive: Empowerment Evaluation involves engaging stakeholders in the evaluation process, promoting their participation and ownership of the evaluation process, and building their capacity to take control of their own development. This makes the process more inclusive and democratic, giving stakeholders a voice and a stake in the outcome.

Empowerment Evaluation promotes learning and improvement: By involving stakeholders in the evaluation process, Empowerment Evaluation promotes learning and improvement, enabling stakeholders to identify areas for improvement, make informed decisions, and take action to address identified needs.

Empowerment Evaluation builds capacity: Empowerment Evaluation builds the capacity of stakeholders to collect and analyze data, make informed decisions, and take action to address identified needs. This enhances the sustainability of development efforts by building local capacity and ownership.

Empowerment Evaluation supports sustainable development: By pro-moting participation and ownership, building capacity, and promoting learning and improvement, Empowerment Evaluation supports sustainable development and positive social change.

Challenges

Time-consuming and resource-intensive: Empowerment Evaluation can be time-consuming and resource-intensive, requiring extensive collaboration, data collection, and analysis. This can be a challenge, particularly in resource-constrained settings.

Requires skilled facilitation: Empowerment Evaluation requires skilled facilitation to ensure tenement of stakeholders in the evaluation process and that the process is participatory and inclusive. This can be a challenge, particularly in settings where there may be power imbalances or where stakeholders may have limited experience in evaluation.

May be resistant to change: Empowerment Evaluation requires stakeholders to be open to change and to be willing to take action to address identified needs. This can be a challenge, particularly in settings where there may be resistance to change or where stakeholders may have competing priorities.

May not be appropriate in all settings: Empowerment Evaluation may not be appropriate in all settings, particularly in situations where there may be safety or security concerns, or where stakeholders may not have the capacity or resources to participate fully in the evaluation process.

Future Directions of Empowerment Evaluation

Empowerment Evaluation has the potential to make a significant contribution to the field of evaluation and to the broader field of social and human development. Here are some future directions for Empowerment Evaluation:

Greater use of technology: Technology offers opportunities for greater engagement and participation of target group in the evaluation process, and for collecting information and analysis. Empowerment Evaluation could benefit from greater use of technology, including online collaboration platforms, mobile data collection tools, and data visualization and analysis software.

Integration with other evaluation approaches: Empowerment Evaluation can be integrated with other evaluation approaches, such as participatory evaluation, developmental evaluation, and feminist evaluation, to enhance its effectiveness and impact.

Advancing the theory of Empowerment Evaluation: Empowerment Evaluation is a relatively new approach to evaluation, and there is much that remains to be understood about its theoretical underpinnings and its relationship to other theories of evaluation and social change. Further research is needed to advance the theory of Empowerment Evaluation.

Emphasis on equity and social justice: Empowerment Evaluation has the potential to promote equity and social justice by giving voice to marginalized and underrepresented groups, building their capacity to participate in the evaluation process, and promoting action to address social and health disparities. Future directions for Empowerment Evaluation could place a greater emphasis on equity and social justice, including the use of anti-oppressive and anti-racist principles in the evaluation process.

Integration with policy and decision-making: Empowerment Evaluation can contribute to policy and decision-making by providing stakeholders with evidence-based information and empowering them to make informed decisions about program and policy development. Future directions for Empowerment Evaluation could place a greater emphasis on integration with policy and decision-making processes.

The future of Empowerment Evaluation is bright, with opportunities for greater use of technology, integration with other evaluation approaches, advancing the theory of Empowerment Evaluation, emphasis on equity and social justice, and integration with policy and decision-making. These future directions have the potential to improve the effectiveness and result of Empowerment Evaluation in promoting positive social change and sustainable development.

Conclusion: Empowering Communities for Sustainable Development

Empowerment Evaluation is an approach to evaluation that aims to empower communities and promote sustainable development. By involving stakeholders in the evaluation process, building their capacity, promoting learning and improvement, and supporting action, Empowerment Evaluation fosters a more democratic and inclusive approach to evaluation and development.

The key principles of Empowerment Evaluation emphasize collaboration, participation, and ownership, and the role of the evaluator is to facilitate the process and build the capacity of stakeholders to take control of their own development. The process is democratic and inclusive, and it recognizes the importance of local knowledge and expertise.

Empowerment Evaluation has been applied in a variety of contexts, including health, education, and community development, and it has shown promise in promoting positive social change and sustainable development. The approach has many advantages, including promoting participation and ownership, building capacity, promoting learning and improvement, and supporting sustainable development.

However, there are also challenges to Empowerment Evaluation, including the need for skilled facilitation, resource constraints, and resistance to change. To address these challenges and maximize the impact of Empowerment Evaluation, future directions could include greater use of technology, integration with other evaluation approaches, emphasis on equity and social justice, and integration with policy and decision-making.

Overall, Empowerment Evaluation has the potential to empower communities and promote sustainable development by fostering a more democratic and inclusive approach to evaluation and development. By building the capacity of stakeholders, promoting learning and improvement, and supporting action, Empowerment Evaluation can contribute to positive social change and sustainable development.

Theory Driven Evaluation

Conceptually, theory-driven evaluations should explicate a program theory or model. Empirically, theory-driven evaluations seek to investigate how programs cause intended or observed outcomes.

It proposes a perspective, called the theory-driven approach, to deal with issues of validity. The central argument is that a model or theory should be formulated in a project assessment and the modelling process should include the identification of potential threats to validity in research.

Correlation between theory and practice of evaluation

Integration, in pedagogy, may mean 'making connections within a major, between fields, between curriculum and co curriculum, or between academic knowledge and practice' Huber *et al.*, (2005).

Theory is different from practice as theory is thinking when practice is doing. However, theory helps us explain knowledge and practice. There are theories to explain what is being observed.

The connection between practice and theory is important as it demonstrates your ability to use evidence to increase your understanding of key concepts, justify your decision making, and inform future practice.

This can be done in coursework through using, for example, authentic classroom materials, videotapes of teaching and learning, cases and by invoking students' own experiences as children and learners in schools." Across the teacher education curriculum active learning models should include several teaching modes.

Evaluation Forms

An evaluation form is a form that is designed primarily to collect feedback on a person, organization, product, or event. Feedback collected using an evaluation form needs to be as insightful and actionable as possible, so your evaluation forms must have all the right fields.

What Is an Evaluation Form?

An evaluation form is a form that is designed primarily to collect feedback on a person, organization, product, or event.

Feedback collected using an evaluation form needs to be as insightful and actionable as possible, so your evaluation forms must have all the right fields.

Employee Evaluation Form

Skill	Very poor	Poor	Neutral	Great
1. Quality of work				
2. Group work				
3. Creativity				
4. Honesty				
5. Initiative				
6. Punctuality				
7. Attendance				
8. Technical ability				

Customer Service Evaluation Form

Please rate your experience with us

Experience	Very poor	Poor	Neutral	Great
1. Helpfulness				
2. Communication				
3. Knowledge				
4. Service efficiency				

Peer Evaluation Form

	Strongly Agree	Agree	Undecided	Disagree	Strongly Disagree
Completes task by due dates					
Makes helpful suggestions					
Communicates effectively					
Collaborates effectively					
Up holds company's value					

Performance Improvement Plan Form

Area of Competency	Tick any one
1. Efficiency	
2. Attendance	
3. Quality	
4. Team work	
5. Productivity	
6. Professional conduct	

Event Evaluation Form

Events	Very Dissat-isfied	Dissatisfied	Neutral	Satisfied	Very Satisfied
1. Venue					
2. Coordinator hosting skill					
3. Event content					
4. Entertainment value					
5. Volunteers' attitude					
6. Cleanliness					
7. Timeliness					

Training Course Feedback Template

Training	Strongly Disagreed	Disagreed	Neutral	Agree	Strongly Agree
1. Learned something valuable					
2. Enjoyed the ice breaker activity					
3. The instructor was encouraging					
4. Printed materials were easy to follow					
5. The role play activity was interesting					

Workshops, conference and apprenticeship / internship

Workshop

Evaluation is a necessary component of all training, including workshops. Evaluation can provide information about the teaching and learning that occur. During a workshop and document the extent to which long-term objectives were achieved after a workshop.

Basic of Evaluation

The three widely used valuation methods used in business valuation include the Asset Approach, the Market Approach, and the Income Approach. The three approaches vary in the way they conclude to value, but the goal of each approach is still the same: to assess the value of the operating entity (i.e., the business).

Starting of workshop

How do I start learning valuation?

Description

1. Understanding the business model.
2. Preparing the financial statements-income statement, balance sheet and cash flows statement.
3. Forecasting the financial statements.
4. Discounted Cash Flows based valuation techniques
5. Sensitivities of valuation estimates.

Evaluation conference

It is a training program for people to help them acquire knowledge about industries. It provides opportunity to gain knowledge and skill about specific industry and specific operation's students also undergo such provision.

3

How to Conduct Evaluation

Identify and Describe Program You Want to Evaluate

The first step in conducting a program evaluation is to clarify why we are doing it, what we want to achieve, and what we will focus on. This will help us to define the evaluation questions, criteria, indicators and methods that will guide our data collection and analysis.

What is Program Evaluation?

Evaluation is a systematic method for collecting, analysing, and using data to examine the effectiveness and efficiency of programs and, as importantly, to contribute to continuous program improvement.

What is the Purpose of Program Evaluation?

Program evaluation is a universal concept used in government, for-profit and non-profit organizations. It serves program development and assists in decision making to improve a product or enhance social welfare services. Being able to monitor the pitfalls and milestones of a project is vital to its success.

Program evaluation is an organized and systematic technique of collecting information regarding a program, sometimes referred to as a project. In this context, a program refers to both the process of delivering a service and producing and distributing a product. The evaluation's aim is to answer specific questions relating to a program's efficiency and to help make informed and timely decisions. The program evaluation articulates what needs to be understood about a program and identifies how to improve its functionality.

This means that when a project has been implemented, the organization should periodically assess if it is on track to achieve its goals. If there is an opportunity to improve, the project manager can make the necessary adjustments and start monitoring the efficacy of the revisions. If there is an improvement, the team can record and document their success for future reference and further development.

Identify the Phases of Program (Design, Start-up, Going, Wrap-up, Follow up)

There are four project life cycle phases: initiation, planning, execution, and closure. And if we monitor each, we can systematize them and understand where there's room for improvement. Especially if we review them separately, instead of just treating all the phases as one big project.

Sl. No.	Phases
1	Initiation in the Project Life Cycle
2	Project Planning
3	Project Execution
4	Project Duration Closure

Initiation in the Project Life Cycle

In the project initiation phase, we're setting the foundations for later success. No other aspect of communication means as much as the communication we will experience at the very beginning. Again, our goal here is to systematize how we process the initial information and make sure we are always getting the information.

The first step is defining the project through:

1. Identifying a need or a problem that the project will solve
2. Identifying opportunities, we can use to solve the problem
3. Understanding whether the project is feasible and will solve the problem
4. Defining the scope of the project and the deliverables
5. Identifying the stakeholders and defining the necessary resources.

After fulfilling these requirements, we will be able to create **a project charter** containing all the information on purposes, objectives, resources, and other aspects.

The analysis we conduct in this phase of the project life cycle will help us understand how our project will be progressing in the future, as well as organize and assemble all the necessary people and resources.

Project Planning

In this part of the project management life cycle, we

i. Set a budget and estimate a timeframe
ii. Establish milestones
iii. Perform a risk analysis
iv. Define tasks and responsibilities
v. Create a workflow

We should look at this phase both strategically and practically. Using a solution that leverages project management templates is a great way to get our projects up and running quickly and with the right amount of process. Try to understand how each task leads to the ultimate goal of solving the problem and keep in mind that the tasks should be practical and easy to accomplish. The most important part of project planning as a stage is definitely **a risk analysis** which can help you identify any potential roadblocks.

Project: Execution

Now that we have set up everything correctly, it's time to bring the team on board and get to work. The execution phase is the one where communication can really make or break a project. During this phase, we should communicate both with our team to make sure they are advancing as predicted, and the stakeholders to keep them updated. We should also establish key performance metrics for project tracking. This way, we will be able to see how everything's going and adjust if needed.

How to Monitor Project Performance?

If we already using Microsoft 365, we can just get an upgrade to project management with Project Central. Our standard tool becomes more comprehensive, without becoming complicated. With Project Central, we will be able to: visually manage our project. It's Time for Closure.

Finally, when the project is done, it's time to evaluate the success of it, the performance of our team, and write a report.

Project Design

The 7 steps of project design.

There are seven steps that make up a successful project design process. These include everything from defining goals and baseline objectives to strengthening our strategy to help us stay organized while managing a new project.

1. Define Project goals
2. Determine out-comes
3. Identify risk and constraints
4. Refine strategy
5. Make Budget estimate
6. Create contingency plan
7. Document mile stone.

Start-Up

Project start up is a very short phase of the project management cycle. It is a transition step between project planning and execution. While it is a short step, important events occur, including: Base lining of the project plan with management sign-off.

On-going Project

Means a multi-year project undertaken by a Company in fulfilment of its CSR obligation having timelines not exceeding three years excluding the financial year in which it was commenced, and shall.

Wrap-up Project

1. Make sure that all words are spelled correctly.
2. Try different keywords.
3. Try more general keywords.
4. Try fewer keywords.

Follow up Project

Follow-up is defined as "the monitoring and evaluation of the impacts of a project or plan for the management of, and communication about the performance of that project or plan".

Determine the feasibility of implementing evaluation

According to White (2000) a feasibility study is an evaluation of the possibility that a particular approach has any potential for success after further research and implementation. Feasibility evaluations provide results of interest to researchers and to sponsors of research.

Implementation feasibility is the framework of facilitating and accelerating the successful implementation of a regional energy plan by evaluating whether the plan at hand is fiscally, technologically, legally, politically, administratively, culturally, and ethically feasible When we face a complex problem, we need to generate and evaluate different solutions. But how do we know which solution is the best one for our situation? How do we compare the merit and demerit of each option? How do we assess the feasibility and impact of our solutions?

Step 1: Define the criteria

The beginning step is to define the criteria that we will use to evaluate our solutions. Criteria are the standards or requirements that our solution must meet to be considered effective and desirable. For example, some common criteria are cost, time, quality, sustainability, and stakeholder satisfaction. We can use a brainstorming technique to generate a list of criteria that are relevant to our problem and context. Then, we can prioritize and select the most important ones to focus on.

Step 2: Score the solutions

The second step is to score the solutions according to the criteria that we have defined. Scoring is a way of quantifying and comparing the performance of each solution on each criterion. We can use a simple numerical scale, such as 1 to 5, where 1 means very low and 5 means very high. Alternatively, we can use a more sophisticated method, such as a decision matrix or a weighted scoring model, where you assign different weights to each criterion based on their importance. The goal is to assign a score to each solution on each criterion and then calculate the total score for each solution.

Step 3: Analyse the trade-offs

The third step is to analyse the trade-offs between the solutions. Trade-offs are the compromises or sacrifices that we have to make when we choose one solution over another. For example, a solution that is cheaper may be less durable, or a solution that is faster may be less accurate. We can use a visual tool, such as a radar chart or a spider diagram, to plot the scores of each solution on each criterion and see how they compare. The goal is to identify the strengths and weaknesses of each solution and the trade-offs that we have to accept.

Step 4: Evaluate the risks

The fourth step is to evaluate the risks associated with each solution. Risks are the uncertainties or potential problems that may arise when we implement our solution. For example, some common risks are technical failures, budget overruns, customer complaints, or legal issues. We can use a risk analysis technique, such as a SWOT analysis or a risk matrix, to identify and assess the likelihood and impact of each risk.

By following these four steps, we can evaluate the feasibility and impact of different solutions and make an informed decision.

Key stakeholders

The key to effective stakeholder consultation is to control the interchange of information using digital channels and tools. These tools are designed not only to communicate but also to isolate the conversations in narrow fields of interest that are pertinent to our project.

What is a Stakeholder?

A stakeholder is anyone who can affect or is affected by the actions of a corporation. The concept of the stakeholder was first used in 1963, at the Stanford Research Institute. It defined stakeholders as those groups without whose support the organization would cease to exist.

It was further developed in the 1980s by R. Edward Freeman and has subsequently been widely accepted in the business community. Broadly speaking, the stakeholder is actually anyone who has an interest in the matter to hand, and the net can be cast as widely, or as tightly, as the business wishes.

Ways of Data Collection (Qualitative, Quantitative, Mixed)

If the researcher uses numbers, they are using a quantitative measure; if they use a descriptive style, it is qualitative measure; and if they are somewhere in between, it is a mixed method.

Quantitative data is measurable numerical data researchers collect by asking close-ended or multiple-choice questions using surveys, polls, questionnaires, and other methods.

Qualitative method of data collection

There are three core methods for data collection in qualitative research interviews, focus groups and observation provide researchers with rich and deep insights. All methods require skill on the part of the researcher, and all produce a large amount of raw data.

If there were a group of people in a room, qualitative data could describe how they feel, what they look like, what clothes they are wearing, or the motivations of why they're here. Whereas quantitative data about the same group may include the number of people in the group, their age, or the temperature in the room.

Qualitative data: Make meaning through methodologies and methods like grounded theory, thematic analysis, content analysis and narrative analysis.

Quantitative data: Make meaning through mathematical techniques such as mean, median standard deviation, linear regression etc.

A mixed methods study combines quantitative and qualitative data collection and analysis in one study. Individually, these approaches can answer different questions, so combining them can provide us with more in-depth findings.

Select Data Collection Techniques (Surveys, Interviews and Questionnaire with Different Types)

Data collection is the process of collecting and evaluating information or data from multiple sources to find answers to research problems, answer questions, evaluate outcomes, and forecast trends and probabilities. It is an essential phase in all types of research, analysis, and decision-making, including that done in the social sciences, business, and healthcare.

What Are the Different Data Collection Methods?

Primary and secondary methods of data collection are two approaches used to gather information for research or analysis purposes. Let's explore each data collection method in detail:

1.	Primary Data Collection	1. Surveys 2. Questionnaires 3. Interviews 4. Observation 5. Experiment 6. Focus Groups
2.	Secondary Data Collection	1. Published sources 2. On line data base 3. Government 4. Instructional record 5. Publicly Available Data 6. Past Research Studies

Primary Data Collection

Primary data collection involves the collection of original data directly from the source or through direct interaction with the respondents. This method allows researchers to obtain firsthand information specifically tailored to their research objectives. There are various techniques for primary data collection, including:

A. Surveys & Questionnaires: Researchers design structured questionnaires or surveys to collect data from individuals or groups. These can be conducted through face-to-face interviews, telephone calls, mail, or online platforms.

B. Interviews: Interviews involve direct interaction between the researcher and the respondent. They can be conducted in person, over the phone, or through video conferencing. Interviews can be structured (with predefined questions), semi-structured (allowing flexibility), or unstructured (more conversational).

C. Observations: Researchers observe and record behaviors, actions, or events in their natural setting. This method is useful for gathering data on human behavior, interactions, or phenomena without direct intervention.

D. Experiments: Experimental studies involve the manipulation of variables to observe their impact on the outcome. Researchers control the conditions and collect data to draw conclusions about cause-and-effect relationships.

E. Focus Groups: Focus groups gather group of individuals who discuss specific topics in a normal setting. This method helps in understanding opinions, perceptions, and experiences shared by the participants.

Secondary Data Collection

Secondary data collection involves using existing data collected by someone else for a purpose different from the original intent. Researchers analyze and interpret this data to extract relevant information. Secondary data can be obtained from various sources, including:

A. Published Sources: Researchers refer to books, academic journals, magazines, newspapers, government reports, and other published materials that contain relevant data.

B. Online Databases: Numerous online databases provide access to a wide range of secondary data, such as research articles, statistical information, economic data, and social surveys.

C. Government and Institutional Records: Government agencies, research institutions, and organizations often maintain databases or records that can be used for research purposes.

D. Publicly Available Data: Data shared by individuals, organizations, or communities on public platforms, websites, or social media can be accessed and utilized for research.

E. Past Research Studies: Previous research studies and their findings can serve as valuable secondary data sources. Researchers can review and analyze the data to gain insights or build upon existing knowledge.

Different types of survey questions for use

Closed-ended questions

Examples

Sl. No.	Statement	Yes	No
1	Are you feeling better today?		
2	May I use the bathroom?		
3	Is the prime rib a special tonight?		
4	Will you please do me a favor		

Open-ended questions

Example

Sl. No.	Questions
1	How would you describe our working culture?
2	What would you do to improve productivity?
3	What would you think if we moved to a four-day working week?
4	How would you feel if we were to change our benefits package?

Multiple choice survey questions

Example

Sl. No.	Reasons of poverty of tribal people	Tick as you feel appropriate
1	Illiteracy	
2	Ignorance	
3	Less attention by Govt	
4	Lack of funds	

Rating survey questions

Example:

a. How likely are you to recommend us to a friend or colleague?

On a scale of 0–10?

How would you rate our customer service on a scale of 1–5?

Likert scale survey questions

Example:

"How pleased are you with your job?" could be a question from the Likert scale examples. The answer could be on a scale of 1 to 5, where 1 is strongly dissatisfied and 5 strongly satisfied with this question. This type of question collects useful data quickly.

Drop-down survey questions

Dropdown survey questions allow questionnaire respondents to choose their answer from a pre-established list of options by clicking on a drop-down menu.

A dropdown question allows respondents to choose an option from a list of options displayed in the dropdown menu. The dropdown menu is visible once the respondent clicks on the down arrow. Depending on the requirements, a dropdown menu can be configured to let respondents select one or multiple options.

Identify Population and Select (Sampling for Evaluation Sample Size, Errors, Sampling Techniques

i) Population

The study population is the subset of the target population available for study (e.g. schizophrenics in the researcher's town). The study sample is the sample chosen from the study population. (i) Population of the study.

ii) Sampling for Evaluation

Devising a thoughtful sampling strategy is one way to ensure that an evaluation is practical and achievable. Sampling is the process of selecting units (i.e., individuals or groups of participants) from a population of interest.

Different types of Sampling

Probability and Non-Probability Sampling Method

Types of Sampling	Definition
1. Simple Random Sampling	Simple random sampling provides opportunity to draw sample from population where each member has equal chance to be selected.
2. Systematic Sampling	In this process the samples are selected at regular intervals till required number is reached.
3. Stratified Sampling	On specific parameters the population is classified to different strata and from each strata required number of samples are drawn following random methods.
4. Cluster Sampling	In a location different clusters are selected based on certain criteria. The clusters are then select adopting random methods. The cluster should be representatives of population.
5. Convenience Sampling	This type of sampling belongs to non-probability sampling method and the units are selected for inclusion for study. This can be due to geographical proximity, availability at a given time, or willingness to participate in the research.
6. Purposive Sampling	Convenience sampling is a non-probability sampling method and the units are selected for inclusion in study because of easiness for the researchers to access. This can be due to geographical proximity, availability at a given time, or willingness to participate in the research.
7. Snowball Sampling	Snowball sampling or chain-referral sampling is defined as a non-probability sampling technique in which the samples have rare traits. In this technique, subjects provide referrals to recruit samples required for investigation.
8. Quota Sampling	Quota sampling is a non-probability sampling method that depends on the non-random selection of a decided number or part of units. This is called a quota.

Sample Size Sampling Techniques

In statistics, the **sample size** is the measure of the individual samples for inclusion in investigation. If we are testing 50 samples of people who watch TV in a city, then the sample size is 50.

Determining Sample Size

1. Determine the population size (if known).
2. Determine the confidence interval.
3. Determine the confidence level.

4. Determine the standard deviation (a standard deviation of 0.5 is a safe choice where the figure is unknown).
5. Convert the confidence level into a Z-Score.

Good Sample Size

Sampling ratio (sample size to population size): Generally speaking, the smaller the population, the larger the sampling ratio needed. For populations under 1,000, a minimum ratio of 30 percent (300 individuals) is advisable to ensure representativeness of the sample.

How to calculate sample size

- **N = Population size**
- **e = Margin of error (percentage in decimal form)**
- **z = z-score**

How to calculate sample size

N=population size • e = Margin of error (percentage in decimal form) • z = z-score

Formula of Sample Size =

$$\frac{\dfrac{z^2 \times P(1-P)y}{e^2}}{1+\dfrac{z^2 \times P(1-P)y}{e^2 N}}$$

The z-score is the number of standard deviations a given proportion is away from the mean. To find the right z-score to use, refer to the table below:

To find the right z-score to use, refer to the table below:

Desired Level of Confidence	Z-Score
80%	1.28
85%	1.44
90%	1.65
95%	1.96
99%	2.58

Things to watch for when calculating sample size

- If we want a smaller margin of error, we must have a larger sample size given the same population.
- The higher the sampling confidence level you want to have, the larger your sample size will need to be.

Statistical Error

Error (statistical error) describes the difference between a value obtained from a data collection process and the 'true' value for the population. The greater the error, the less representative the data are of the population. Data can be affected by two types of error: sampling error and non-sampling error.

Sampling Techniques

Sampling is a technique of selecting individual members or a subset of the population to make statistical inferences from them and estimate the characteristics of the whole population.

Collect, analyses and explanation of qualitative and quantitative Data, for Evaluation Data Analysis)

Data Collection

Data collection is the method of gathering and analysing information on relevant variables in a predetermined, methodical way so that one can respond to specific research questions, test hypotheses, and assess results.

Data Collection Methods

i) Surveys, quizzes, and questionnaires.
ii) Interviews.
iii) Focus groups.
iv) Direct observations.
v) Documents and records (and other types of secondary data, which won't be our main focus here.

Data Analysis

Data Analysis is the process of systematically applying statistical and/or logical techniques to describe and illustrate, condense and recap, and evaluate data.

Four Types of Data Analysis

i) Descriptive Analysis
ii) Diagnostic Analysis
iii) Predictive Analysis
iv) Prescriptive Analysis

Data Interpretation

Data interpretation is the process of reviewing data and arriving at relevant conclusions using various analytical research methods. Data analysis assists researchers in categorizing, manipulating, and summarizing data to answer critical questions.

Here are some steps to interpreting data correctly.

Gather the Data

The very first step in data interpretation is gathering all relevant data. We can do this by first visualizing it in a bar, graph, or pie chart. This step aims to analyse the data accurately and without bias. Now is the time to recall how we conducted research.

Develop Discoveries

This is a summary of our findings. Here, we thoroughly examine the data to identify trends, patterns, or behaviour. If we are researching a group of people using a sample population, this is the section where we examine behavioural patterns. We can compare these deductions to previous data sets, similar data sets, or general hypotheses in our industry. This step's goal is to compare these deductions before drawing any conclusions.

Draw Conclusions

After we have developed our findings from our data sets, we can draw conclusions based on our discovered trends. The findings should address the questions that prompted our research. If they do not respond, inquire about why; it may produce additional research or questions.

Give Recommendations

The interpretation procedure of data comes to a close with this stage. Every research conclusion must include a recommendation. As recommendations are a summary of our findings and conclusions in brief. There are only two options for recommendations; we can either recommend a course of action or suggest additional research.

Communicate Findings (Reporting Plan, Evaluation Report Types, Reporting Results, Reporting the Tips, Reporting Negative Findings

Communicate Findings

Communication of research findings is the final and one of the most important steps of research process. Research communication is defined as the process of interpreting or translating complex research findings into a language, format and context that non expert can understand.

Reporting Plan

Business reporting is an essential part of any planning process in the workplace as it consists of providing data and information to specific audiences. Examples of business reports include financial plans, customer service reviews and marketing research results.

Report of Evaluation

It is important to know the types of evaluation that can be conducted over a program's life-cycle and when to use. The important ones are process, impact, outcome and summative evaluation.

Reporting results involves communicating the findings of our evaluation in a manner that makes sense to your target audience(s). Although the typical way to disseminate results is through an evaluation report, there are other means that may be more accessible and useful to a particular stakeholder group.

Reporting the tips

The report should include the following elements:

1. Title
2. Date of report
3. Name and the name of the company colleagues who accompanied you (if any)
4. Names of clients or partners we have seen during the trip
5. Exact dates of each of the events of the trip accompanied by their descriptions
6. Summary
7. Expenses and earnings

Reporting Negative Findings

Researchers may also consider uploading their findings to a free database such as 'fig share'. However, publishing in peer-reviewed, widely accepted journals still remains the gold standard! An article describing negative findings will be considered credible only when it is published in a reputed peer-reviewed journal.

Apply and Use Findings (program continuation/discontinuation improve ongoing program, plan future program and inform program stakeholders)

Program Improvement

1. Compare outcomes with the program objectives and goals.
2. If outcomes are below expectations, find alternatives to improve the program
3. Review strengths and participants' comments to identify alternative approaches to address weaknesses.
4. Identify or develop alternatives before the next program cycle.

Use of our Evaluation Findings

Accountability

Meeting financial education needs of diverse societal groups is a challenging task because of limited resources. Therefore, financial educators must be prepared to distinguish effective financial education programs from ineffective ones. This determination can be achieved only if the educator uses program evaluation as a measure to discern effective programs. If the financial education program is effective in terms of bringing desired outcomes to the target population, the educators should continue to implement the program. Additionally, evaluation is useful as a public accountability measure to secure stakeholder support for effective financial education programs. Communication of evaluation findings with stakeholders such as funding agencies, decision makers and policymakers is essential to convince them about the importance of investing in financial education. Financial education evaluation provides a means to receive due attention and garner support from the public and policymakers.

Funding

Funding is essential to continue financial education programs. Funding agencies look for successful programs and are sensitive to cost effectiveness. Therefore, it is important to ensure the cost effectiveness of financial education programs. Cost effectiveness can be illustrated using evaluation data to distinguish effective programs from ineffective programs. Competition for limited funding is very high. Therefore, the educator must be prepared to present strong financial education programs as examples when applying for funding. Follow these important steps when presenting evaluation data to support continuous funding.

i) Make evaluation an integral part of the program.

ii) Document participant reactions (e.g., satisfaction, relevance to their lives).

iii) Document evaluation findings about program quality to demonstrate strong implementation. Clearly communicate program outcomes with the funding agency.

iv) Communicate outcomes with potential funding agencies to seek future funding.

Communication with Stakeholders

Partnerships involve two or more individuals, groups or organizations working together for a common goal. When two or more partners work together, it is important that they mutually understand the program development and delivery process. If changes are made to program delivery, all the partners should understand the rationale for those changes. Formative evaluation data can be utilized to justify the changes needed without being biased to one partner. This is an essential element to managing partnerships.

In conclusion the program, partners and stakeholders normally like to share credit for the program and be informed about the program's successes. The following tips are helpful when using evaluation results for building strong partnerships:

i) Keep partners informed about the evaluation plan.

ii) Share ongoing evaluation data about program quality with partners on a regular basis.

iii) Document and share outcomes with partners to highlight the worth of the partnership.

iv) Acknowledge the contribution of each partner.

Marketing

Evaluation findings also can be useful for marketing the program to potential participants or the broader community. Documenting improvements in financial education made by participants or participants' positive perceptions of the program can be persuasive for program marketing materials.

Program Improvement

Evaluation data can be used to inform program improvement efforts. Reviewing the evaluation findings helps the educator decide whether the program is achieving set objectives or not. If the program is not being implemented in a high-quality manner or the program outcomes are below expectation, the educator needs to identify alternatives to modify the program before presenting it next time. Program improvement can be done only if the educator has collected formative evaluation data to identify strengths and weaknesses. If the strengths and weaknesses of the program have been identified, the educator can plan alternatives to eliminate weaknesses and capitalize on strengths to make the program stronger next time (or to conduct mid-course corrections).

Use these helpful tips to use evaluation data for the improvement of educational programs:

i) Compare outcomes with the program objectives and goals.

ii) If outcomes are below expectations, find alternatives to improve the program.

iii) Review strengths and participants' comments to identify alternative approaches to address weaknesses.

iv) Identify or develop alternatives before the next program cycle.

v) Review process evaluation data such as participants' ratings of instructors and education materials.

vi) When there are low ratings of educational materials or instructors, modify that item or find alternatives before presenting the next program.

vii) Perform continuous evaluation to further improve the program. This process maximizes the cost-effectiveness of financial education.

4

Evaluating the Evaluation

Evaluating the Evaluation

To present clarity, representation of components of program, stake holders, senility, needs of target group, sample, data, technical adequacy, data collection methods cost, recommendation the steps have been described below as per syllabus contents prescribed by ICAR.

Evaluating the Evaluation

1. To determine or fix the value of evaluation.
2. To determine the significance, worth, or condition of usually by careful appraisal and study.

Representation of program components and stakeholders

Program Components

i) Who (Target audience)
ii) Where (Setting)
iii) What (Content, or approach and activities)
iv) How (Mode of delivery)

Components of a Program

1. Policy Procedure
2. Mission and Goal
3. Organization Structure
4. Diversity and inclusiveness
5. Assessment and Evaluation
6. Training and Development
7. Tools and Technology

Stakeholders

The individuals, organizations actively participating in program are the stakeholders. Their actions, reactions may e positive or negative affect the programs. Such is the definition of stake holders that we conceive in this chapter. This includes customers, users, suppliers, and investors. The, stake-holders don't always work for the project manager.

Sensitivity

The main goal of the evaluation process is to see if the programme is having an impact on the social problem it is trying to solve; as a result, the measurement instrument must be sensitive enough to detect these potential changes (Rossi *et al.*, 2022).

Representative of Needs in Program Evaluation

To create a fully informed plan for an effective program, it is essential to know what the problems are that need to be addressed (e.g., where the gaps in community preparedness are, who is most at risk, where there are limited resources to improve preparedness) and the resources that are available. A need assessment is the process of gathering information about the current conditions of a targeted area, or the underlying need for a program. A resources assessment is the process of gathering information about the resources available to address a particular need or risk.

The needs and resources assessment of our community and target population can help identify the most prevalent community risks, hazards and vulnerabilities, gaps, conditions that call for change, and what community resources are available to assist our efforts. Knowing current needs will help with setting realistic goals and desired outcomes.

Complete this step by

1. Gathering information on the problems or needs in your community
2. Identifying existing resources that address these needs
3. Prioritizing needs to select those you can address
4. Specifying your target population

Tools Used in This Step

The Data Catlog Tools helps to determine which data source wish to use and who will be responsible for collection of data. If Public Health Department has conducted hazards and vulnerability analysis this may serve as an important data source because it contains information about community risk and vulnerable population.

Sample and Data, Technical Adequacy Dealt under Unit 1 of Block 1

Scores or information generated from an instrument that is "technically adequate" can reliably and accurately predict performance or behaviour for any given population of interest. In other words, the instrument should be standardized for the population it will be used on, reliable, and accurate in its identification of children who are and who are not at risk. Make no mistake: this is a high but reasonable standard to achieve. In-depth guidelines can be found in Educational and Psychological Testing.

Costs, Recommendations and Reports, Cost what is Cost Evaluation?

1. Definition of Cost Evaluation

Cost evaluation is the process of determining the costs and benefits of an investment or other decision. The costs of an investment can be of both monetary, non-monetary costs and later one is environmental impact. The benefits of investment are financial and non-financial depending on benefits that people expect or receive. Those are increase in productivity or services.

Cost evaluation is a complex and important task. There are several different types of costs that need to be considered, and each decision requires a different approach. Cost evaluation can be divided into two main categories: financial and non-financial. Financial costs are those that involve money, such as capital or labour. Non-financial costs are those that don't involve money, such as the cost of pollution or the cost of risk.

Financial costs can be divided into two main categories: direct and indirect. Direct costs are those that involve spending money itself, such as paying for capital or hiring labour. The other one costs are those that involve spending money on something else, like the cost of fuel or raw materials.

Purpose of Cost Evaluation

Cost evaluation is a process that organizations use to identify and quantify the costs associated with various decisions. The purpose of cost evaluation is to ensure that the decisions made are economically sound and that the benefits of those decisions are worth the costs incurred.

The types of costs that organizations may need to evaluate:

1. Direct costs
2. Indirect costs
3. Opportunity costs
4. Net benefits

Types of Cost Evaluation There are many different types of cost evaluation, some of which are discussed below.

The most common type of cost evaluation

1. Financial Cost: Financial evaluations involve estimating the costs and benefits of a proposed action, and then calculating the difference between the two. Financial evaluations can help identify which actions are the most cost-effective and make sure that the benefits of the proposed action outweigh its costs.

2. Economic Cost: Another type of cost evaluation is economic. Economic evaluations try to measure the impact of a proposed action on society as a whole.
3. Environmental Cost: Environmental evaluations look at the impact of a proposed action on the environment. They can help identify which actions are most harmful to the environment, and help decide which actions are worth taking.

Benefits of Cost Evaluation

Cost evaluation is a process that helps to identify and assess the costs associated with a proposed project or activity. In doing so, it can help to ensure that the benefits of the project or activity are proportionate to its costs.

Sl. No.	Benefits
1	Cost evaluation can help to identify and assess the costs associated with a proposed project or activity.
2	Cost evaluation can help to identify potential savings or opportunities for improvement.
3	Cost evaluation can help to improve decision-making by providing information on both the costs and benefits of different options
4	Cost evaluation can help to ensure that the project or activity is correctly funded
5	Cost evaluation can assist to improve communication between different parties involved in a project or activity.
6	Cost evaluation can help to improve accountability and transparency within a project or activity.
7	Cost evaluation can bring better quality of a project or activity.
8	Cost evaluation can ensure to improve morale within a project or activity.
9	Cost evaluation assures to improve communication between different parties involved in a project or activity.

Steps in calculating cost benefit

There are a number of steps involved in undertaking a cost evaluation. Often these steps are executed in a sequential manner, but sometimes it is useful to think of them as a continuum.

Sl. No.	Steps	Practice
1	Problem Identification	In problem identification, the cost evaluation is initially conceived as a way to resolve a specific issue or problem. This might be the result of feedback from stakeholders or it might be the initiative of a single person or organization.
2	Systematic Analysis	Once the problem has been identified, the next step is systematic analysis. This involves identifying all of the cost elements that might be impacted by the resolution of the problem.

3	Identification of Primary Costs	Once all of the cost elements have been identified, the next step is to determine which ones are primary
4	Assessment of Alternatives	Once primary costs have been identified, the next step is to assess all of the available alternatives. This involves looking at both short- and long-term options and determining which ones best resolve the problem.
5	Selection of an Alternative	Once all of the alternatives have been assessed, the decision regarding which one to select must be made. This can be a difficult task, as it requires consideration of both the impact of each alternative on primary costs and on stakeholders' preferences and expectations
6	Implementation	Once an alternative has been selected, implementation must be planned and executed. This includes designing and implementing changes to systems and processes, training employees, and so
7	Monitoring and Evaluation	After an alternative has been implemented, it must be monitored and evaluated to see how it has impacted primary costs and stakeholders' expectations

Recommendations in Evaluation

At the end of all exercises on evaluation recommendation is given for action leading to benefits. The suggestions for action are based on facts derived out of investigation.

At the end the results of evaluations are to be examined in relation program design, program contents, monitoring, and involvement of managers and stakeholders. It also indicates he action components at different stages of evaluation steps.

Recordation should contain guidance for all those involved. To have beater performance of the programs better decision could be taken based on recommendations. of evaluation. Besides the report also reveals weakness and strength. Therefore, recommendation is the essential ingredient of evaluation programs.

Components of Recommendation

1. Sound program design
2.. Enhance implementation
3. Sound approach in monitoring and evaluation
4. Adequate coverage of program
5. High rate of sustainability of program

Mapping of the Evaluation Findings

Mapping of literature of evaluation indicated thorough examination of previous evaluation studies in relation to design, implementation and benefits.

Quantitative and qualitative data analysis require special care to ensure accurate result. Data transfer, analysis and interpretation can be better if review of literatures are carefully examined. Such attempt also enriches the knowledge of experts involved. It is an important component in scientific investigation about outcomes of programs and projects.

Contextual consideration

The evaluation of any projects needs examination of different variables of the location and people. Such consideration includes social, economic, political, cultural and general behaviour of the residents. Since projects are meant to benefit them, it is essential to analyse the environment of living and livelihood. Society, social norms, communication pattern, interrelationship among various target group should be reflected in the report of evaluation to present complete view of the projects and their effectiveness ion living pattern.

Reflection of Merits and Demerits

We emphasize to bring out strength and weakness of the projects on operation. Such aspects are positive and negative sides in relation to population to be served. The positive aspects are income, profit, sound environment and good social climate. All these factors will lead to development and prosperity leading to think about similar projects.

Adoptable recommendation

The recommendations should be relevant, feasible, and likely to lead to improvements in program performance. The projects meant for people and environment should be adoptable. In other words what can be done should be the theme of outcome of the evaluation exercise.

Specific, measurable, achievable, relevant, and time-bound (SMART) are the symbol of good evaluation report.

Diffusion of result

The outcomes are to be diffused to all concern so that a sensation can be created. The stake holders will know all about projects in the terms of merit and demerits. The program manager, agents, officials and politicians should have detail information about the project including design, implementation, involvement and sharing of benefits.

End Point

The end point is conclusion. It should reveal output clearly in terms of evidence, comprehensive manner pointing out what to be done and what should be avoided. Valuable insights are to be clearly spelled out.

Evaluation Report

An evaluation report is a paper that examines whether a product, service, or process is working, according to a set of standards. It includes an introduction, background information, criteria, evaluation, conclusions, and recommendation.

Structure of Evaluation Report

The structure of the report can vary as per requirements and preferences of the stakeholders, but typically it includes the following sections:

Executive Summary: A brief writings of the evaluation findings, conclusions, and recommendations.

Introduction: An overview of the work context, scope, purpose, and methodology.

Background: Conclusion of the programme or initiative that is being assessed, including its goals, activities, and intended audience(s).

Evaluation Questions: A list of the evaluation questions that guided the data collection and analysis.

Methodology: A description of the information collection methods used in the study like evaluation, including the sampling strategy, data sources, and data analysis techniques.

Findings: A presentation of the evaluation findings, organized according to the evaluation questions.

Conclusions: The chapter should contain the main findings and conclusions, including a determinant of the program or project's effectiveness, efficiency, and sustainability.

Recommendations: A list of specific recommendations for program or project improvements based on the evaluation findings and conclusions.

Lessons Learned: A discussion of the key lessons learned from the tasks that could be applied to similar programs or projects in times to come.

Limitations: A discussion of the factors of limitation of the evaluation, including any challenges or constraints encountered during the data collection and analysis are to properly document.

References: A list of references cited in the evaluation report.

Appendices: Additional information, such as detailed data tables, graphs, or maps, that support the findings and conclusions.

The structure of the report should be clear, logical, and easy to follow, with headings and subheadings used to organize the content and facilitate navigation.

In addition, the presentation of data may be made more engaging and understandable by the use of visual aids such as graphs and charts.

Common Challenges in Writing an Evaluation Report

Writing an evaluation report can be a challenging task, even for experienced evaluators. Here are some common challenges that evaluators may encounter when writing an evaluation report:

Data limitations: One of the biggest challenges in writing an evaluation report is dealing with data limitations. Evaluators may find that the data they collected is incomplete, inaccurate, or difficult to interpret, making it challenging to draw meaningful conclusions.

Stakeholder disagreements: Another common challenge is stakeholder disagreements over the evaluation's findings and recommendations. Stakeholders may have different opinions about the program's effectiveness or the best course of action to improve program outcomes.

Technical writing skills: Evaluators may struggle with technical writing skills, which are essential for presenting complex evaluation findings in a clear and concise manner. Writing skills are particularly important when presenting statistical data or other technical information.

Time constraints: Evaluators may face time constraints when writing evaluation reports, time for action.

Communication barriers: Evaluators may encounter communication barriers when working with stakeholders who speak different languages or have different cultural backgrounds.

By being aware of these common challenges, evaluators can take steps to address them and produce evaluation reports that are clear, accurate, and actionable. This may involve developing data collection and analysis plans that account for potential data limitations, engaging stakeholder.

5

SWOT Analysis and Bar Charts

SWOT Analysis-Concept

Concept of SWOT

The concept of SWOT surfaced in 1960 as, management tool indicating four dimensions of strength, weakness, opportunity and threat and is widely used in business organizations. It's use is very common in judging the prospectus of any organization particularly rural projects. It's four attributes have wider applicability to decide whether to move ahead or not.

Origin of SWOT

The SWOT seeds began to germinate in the early 1950s as an era known that is the war on bigness between the US Government and large corporations from 1948 to 1950. The empirical basis of SWOT started in 1952 within the Lockheed's Corporate Development Planning Department. One of its thus far unknown pioneers, Robert Franklin Stewart, became the head of the Theory and Practice of Planning group at the Stanford Research Institute in 1962. In 1965, Stewart published the so-called SOFT Approach in a report that was used by many large companies worldwide. In it, he presented a logical set of steps (the so-called chain of reasoning) for corporate aim setting. These value judgments are then transferred into senior management's direction-giving statements of corporate purpose. In 1967 The original concept SOFT evolved into: Strengths, Weaknesses, Opportunities and Threats (SWOT Analysis).

SWOT as a Program Management Tool

In this context we shall deal with the following contents.

i) Tools and Techniques
ii) SWOT Analysis Diagram
iii) RACI
iv) Stakeholder Matrix
v) Cause-Effect Diagram
vi) Risk Map and
vii) Radar Chart

Tools and Techniques

Certain tools and techniques used in project management are outlined below.

The internal variables are strength and weakness whereas external variables are opportunity and threat.

SWOT Analysis Diagram

A SWOT analysis can be used to draw out the threats and opportunities facing a programme or project. It has the advantage of being quick to implement and is readily understood. Analysis of the four components brings together the results of internal business analysis and external environmental analysis. Common and beneficial applications of SWOT are gaining a greater understanding and insight into competitors and market position.

A similar and related form of analysis is known as PEST, examining Political, Economic, Social and Technological factors.

RACI - Responsible, Accountable, Consulted, Informed.

Table Showing Task with those Responsible, Accountable, Consulted and Informed

	Role 1	Role 2	Role 3	Role 4	Role 5	Role 6
TASK A	C	C	I			A R
TASK B	A		C	I	R	
TASK C			A	R		I
TASK D	R	C		A		
TASK E			R	A		
TASK F	A	R			C	I

A RACI diagram reveals ructions and performance of the participants in a business or project activity in terms of producing predetermined deliverables. RACI is an acronym formed from the four participatory roles which are:

1. Responsible	Those who undertake the activity or the resources.
2. Accountable	Those who take the credit for success or accountability for failure or the activity manager; and there must be at least one for each activity.
3. Consulted	Those whose opinions are sought
4. Informed	Those who are kept advised of progress

Stakeholder Matrix

A stakeholder matrix is used to map stakeholders in terms of their importance and potential impact on programme or project activity. Stakeholders are the individuals or groups who will be affected by an activity, programme or project. They could include:

1. Senior managers whose business areas are directly or indirectly involved.
2. End-users including customers outside the organisation.
3. Suppliers and partners.

Effective management of the stakeholders' interests includes the resolution of conflicting objectives and representation of end users who may not be directly involved in the activity. Stakeholders' interests can be managed through stakeholder meetings and specific user panels providing input to a requirement specification. The key objective is to capture, align, record, sign off and deliver stakeholder objectives. One way of prioritising this activity is to use a stakeholder matrix.

Power ↑	**Interest** →	
	Low	High
High	Keep Satisfied	Key Players
Low	Minimal Effort	Keep Informed

Risk Map

IMPACT						
	Very High					
	High					
	Medium					
	Low					
	Very Low					
Likely Hood		Very low	Low	Medium	High	Very High

This is a simple representation of **risk** in terms of likelihood and impact. It requires that the probability of a risk occurring is classified as low, medium or high, with a similar classification for the impact if the risk materialises.

A combined risk classification of high probability and high impact if the risk occurs is clearly an important risk. The classification can be extended to include very low and very high.

Risk Profile

A summary risk profile is a simple mechanism to increase the visibility of risks. It is a graphical representation of information normally found on an existing risk register. In some industry sectors it is referred to as a risk map.

The project manager or risk manager needs to update the risk register on a regular basis and then regenerate the graph, showing risks in terms of probability and impact with the effects of mitigating action taken into account. It is essential for the graph to reflect current information as documented in the risk register. The profile must be used with extreme care and should not mislead the reader. If an activity has over 200 risks it will be impractical to illustrate all of the risks. It will be more appropriate to illustrate the top 20 risks, for example, making it clear what is and is not illustrated.

A key feature is the risk tolerance line. It shows the overall level of risk that the organisation is prepared to tolerate in a given situation. If exposure to risk is above this line, managers can see that they must take prompt action. Setting the risk tolerance line is a task for experienced risk managers. It reflects the organisation's attitudes to risk in general and to a specific set of risks within a particular project. The parameters of the risk tolerance line should be agreed at the outset of an activity and regularly reviewed.

Conducting SWOT Analysis

Before starting SWOT, analysis there is need to find answers of the following questions.

1. Status of business
2. Competitors
3. Internal strengths
4. External strengths
5. Prospectus of business with present strength
6. Skills and training required to meet the challenge?

How to Complete Analysis

The analysis can provide strategic insights that enable to improve business operations.

1. Put a team together: Keep people of diverse fields relating to business.
2. Set a goal for SWOT means SWOT for what.
3. Make a list of attributes relating to strengths, weaknesses, opportunities, and threats of the project.

Internal factors are strengths and weaknesses while external factors are opportunity and threat.

Our opportunities and threats may be more difficult to distinguish because we don't think about them daily. Consider things happening outside our company that are just as important as our strengths and weaknesses because they affect us.

4. Refine, organize, and prioritize the ideas in each category

   Once we have identified the internal and external factors of our SWOT analysis, we can start refining and organizing the ideas in each category on our template. Prioritize what's most important to us, the items on our list that we want to tackle first. However, we should also consider items that can be done easily instead of putting them off to form an action plan for more important business initiatives.

5. Create an action plan to address SWOT analysis priorities

   After prioritizing SWOT findings, we can start to brainstorm action plans to align our business goals with four dimensions Take each item out of it and begin creating strategies.

When to conduct analysis

We can conduct analysis at any point to identify these key areas of our business. However, we should always use analysis before making a decision on something that can affect the business considering new initiatives.

When to conduct SWOT analysis

1. Trying a new stream of revenue
2. Changing the business model
3. Changing internal policies
4. Considering new opportunities
5. Forming new partnerships with other companies
6. Company acquisitions
7. Changing strategies in different departments of the business

SWOT analyses are used to help to make better decisions while understanding critical information and data about our business. Doing a SWOT analysis on a quarterly basis can help us to identify new opportunities, threats, and even weaknesses as your business evolves.

Importance of Analysis

It is important for every business because it can help to solve major challenges and answer important questions and making decision easier for leaders. Benefits of a good SWOT analysis include:

1. Visualizing complex problems
2. Knowing of external factors
3. Strengthening business or strategy
4. Easy to prepare SWOT analyses on the part of HR
5. Improved communication

The four components of a SWOT analysis are:

1. Strengths
2. Weaknesses
3. Opportunities
4. Threats

Common questions in SWOT

Strengths:	**Weaknesses:**
1. What do we do best? 2. What unique knowledge, talent, or resources do we have? 3. What advantages do we have? 4. What do other people say we do well? 5. What resources do we have available? 6. What is our greatest achievement?	1. What could we improve? 2. What knowledge, talent, skills and/or resources are we lacking? 3. What disadvantages do we have? 4. What do other people say we don't do well? 5. In what areas do we need more training? 6. What customer complaints have we had about our service?
Opportunities: 1. How can we turn our strengths into opportunities? 2. How can we turn our weaknesses into opportunities? 3. Is there a need in our agency that no one is meeting? 4. What could we do today that isn't being done? 5. How is our field changing? How can we take advantage of those changes? 6. Who could we support? How could we support them?	**Threats:** 1. What obstacles do we face? 2. Could any of our weaknesses prevent our unit from meeting our goals? 3. Who and/or what might cause us problems in the future? How? 4. Are there any standards, policies, and/or legislation changing that might negatively impact us? 5. Are we competing with others to provide service? 6. Are there changes in our field or in technology that could threaten our success?

Advantages and Disadvantages

Advantages	**Disadvantages**
1. Simple and Straightforward Process	1. Unpredictable
2. Offer Multi-Level Analysis	2. Time-Consuming Process
3. Encourages Strategic Planning	3. High Cost

Bar Charts (Gantt Charts and Milestone Charts Bar Charts)

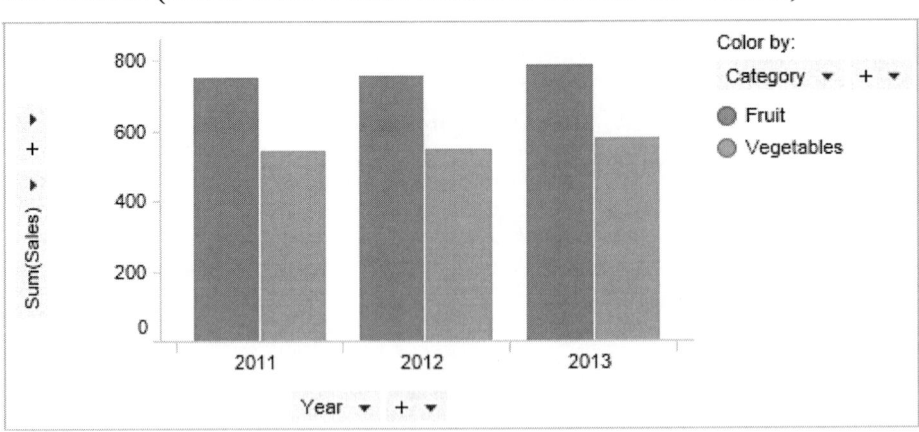

The bar chart displays data using a number of bars, each representing a particular category. The height of each bar is proportional to a specific aggregation (for example the sum of the values in the category it represents). The categories could be something like an age group or a geographical location.

Use of a Bar Chart

Bar charts enable us to compare numerical values like integers and percentages. They use the length of each bar to represent the value of each variable. For example, bar charts show variations in categories or subcategories scaling width or height across simple, spaced bars, or rectangles.

Gantt Charts

It describes graphical presentation of work schedules. It indicates start and end. The items like resources, planning and dependencies are clearly indicated in the chart. The Engineer Haenery Gnatt developed the concept. He added dead line of the time for competition of the project. The terms like milestone date of task and dependency tasks are shown in chart.

Full form of Gantt chart

	Task 1	**Task 2**		**Task 3**	**Task 4**	**Task 5**	**Task 6**	**Task 7**
Task 1								
Task 2								
Task 3								
Task 4								
Task 5								
			Today					
How to use a Gnatt Chart for Project Management								

Generalized Activity Normalization Time Table (GANTT) chart is type of chart in which series of horizontal lines are present that show the amount of work done or production completed in given period of time in relation to amount planned for those projects. It is horizontal bar chart developed by Henry.

The four main components of a Gantt chart are (i) tasks, (ii) timelines, (iii) dependencies, and (iv) progress. Tasks are the individual activities required to complete the project, while the timeline shows when each task will take place.

Important Points

i) Scheduling, managing, and monitoring specific tasks and resources are shown in a project.

ii) It consists of a list of tasks and bars depicting each task's progress.

iii) Different lengths in size and horizontal way it represents the project timeline, which reveal and the start and end dates for each task.

iv) Widely used chart in project management is Gnatt chart.

v) Gantt charts are used in heavy industries for projects like road, houses dams, bridges, and highways.

vi) **Gantt Chart and it's benefits**

It shows slack time, additional time requirement, on critical activities that makes delay in execution.

Milestone Charts

A milestone chart is a horizontal chart that marks the most important steps of our project. Each milestone our team achieves brings closer to completing the project. Besides keeping our team motivated, a milestone chart is also a great way to show project stakeholders how our project is progressing.

What is difference between Gantt chart and milestone chart?

The focus of a Gantt chart is on information such as task durations, dependencies, and resources, while a milestone chart focuses on key dates or achievements. Gantt charts are used for planning, managing, and tracking projects. Milestone charts are for communicating progress and identifying potential issues.

What is the format of a milestone chart?

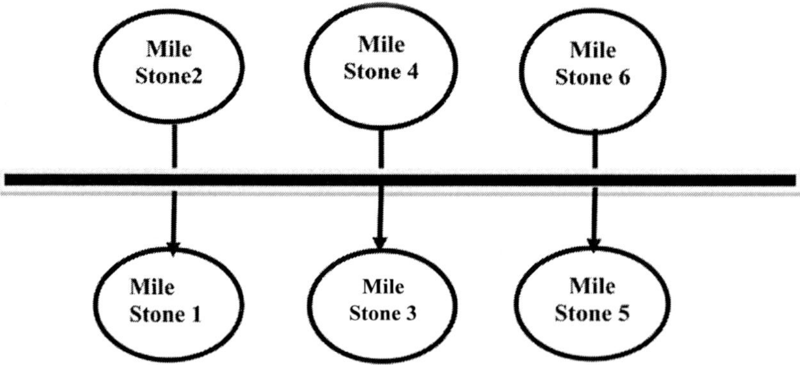

Typical format of a milestone chart

The usual format of a milestone chart is that of a timeline that uses various symbols (squares, diamonds, circles etc.) to separate a project schedule into major phases.

The types of developmental milestones include:

i) Physical

ii) Cognitive

iii) Social

iv) Emotional

v) Communication and language

Bar Charts

A bar chart or bar graph is a chart or graph that presents categorical data with rectangular bars with heights or lengths proportional to the values that

|Usage • Variable-width (verified) • Advantages and Limitations

What is bar chart and types?

Bar graphs are majorly used to compare various observations based on a certain parameters. Considering structure and parameters

The bar graph is classified into the following six types.

1. Horizontal bar graph.
2. Vertical bar graph.
3. Double bar graph (Grouped bar graph)
4. Multiple bar graph (Grouped bar graph)
5. Stacked bar graph.
6. Bar line graph

Characteristics, advantages and limitations

Characteristics

i) SWOT stands for Strengths, Weaknesses, Opportunities, and Threats.

ii) SWOT analysis is a technique for assessing these four aspects of our business.

iii) SWOT Analysis is a tool that can help to analyse what our company does best now, and to devise a successful strategy for the future.

Advantages

i) It's a simple four box framework.

ii) It facilitates an idea of the strengths and weaknesses of the organisation.

iii) It encourages the development of strategic thinking.

iv) It enables authorities to focus on strengths and build opportunities.

Limitations

i) Some SWOT analysis make wrong decision not knowing correctly the bars

ii) It's easy to use wrong data.

iii) The risk of taking much data which lead to 'paralysis by analysis'.

iv) Wrong perception may be some times found misleading managers.

6

Networks

Network in Project Management

A project network is a graph that shows the activities, duration, and inter dependencies of tasks within our project.

Networking is a crucial aspect of project management that can provide various benefits, including access to new job opportunities, industry insights, and professional development opportunities.

What is network with diagram?

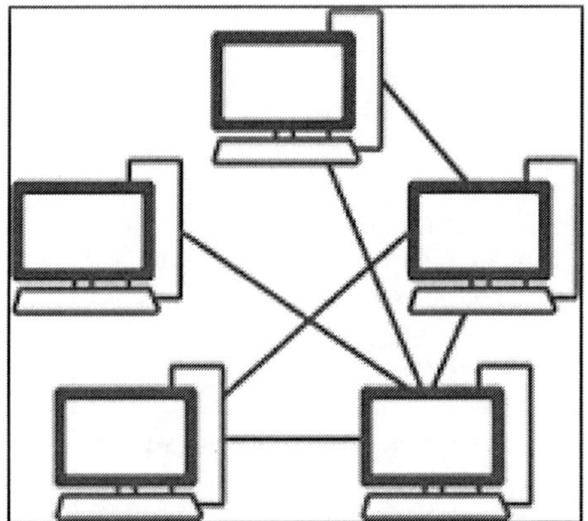

Network and its Origin

A Brief History of Project Management

In today's fast-paced business world, the need for successful project management has become a necessity rather than a luxury. Many tend to think of project management as a new venture for growth and development.

However, project management has been around for thousands of years and was involved in the planning, coordination and construction of the Ancient Wonders of the World.

Today's project management has grown to include industries. Many project management organizations have also taken to adopting the six key drivers of success mentioned in this e-Book.

1. Ancient to 18th Century: Elements of Project Management Transcend Time	Throughout the history of project management, the basic principles of project management have always remained the same. This includes managing resources, maintaining schedules, and coordinating different activities and tasks. Important difference between the ancient marvels of project management and modern-day projects is the ancient marvels did not routinely involve schedule optimization. Many of today's most affluent companies have been able to master these age-old principles of project management by adopting these six crucial elements of successful project management.
2. Project Management Grows in the 19th Century	In the late 19th century, the need for more structure in construction, manufacturing, and transportation sectors gave rise to the modern project management tactics we use today.
3. 1900 to 1950: The Birth of Modern Project	As the 19th century progressed, business leaders began to face the challenges of labour laws and regulations from the federal government. Henry Gantt is known as founding father of modern project management. He developed planning and control techniques to help business leaders succeed and comply with these new regulations. One example is the creation of the famous **Gantt Chart** to ensure monitoring and control of the project schedule. This basic bar chart shows the phases of a project from inception to completion. In 1911 Frederic Taylor published a book, "The Principles of Scientific Management," which was based on his experience in the steel industry. The goal of the book was to give unskilled workers the opportunity to work on new, complex projects by learning skills rapidly through simplicity.
4. 1911: Frederic Taylor	In addition, he identified how many workers would routinely work below capacity through soldiering to ensure future job security. He also identified the need to create incentive-based wage systems and take advantage of time-saving techniques. Many of the principles in Taylor's book align with the success drivers mentioned in this eBook and are still used by companies today.
5. 1950 to 1980s: PERT and CPM	After 2nd world war, project managers began to follow two mathematical ways of conducting and managing projects. **Program Evaluation Review Technique**, or PERT, analyses individual tasks by asserting a minimum amount of time for completion. **The Critical Path Method**, or CPM, factors in a project's activities, how long the activities will take to complete, and the relationship between the activities and their end points. However, CPM quickly became riddled with confusion.

6. 1980 to 2000: Computers and Project Management	The rise of the computer played a major role in the history of project management. Computers brought connectivity and communication to the forefront of project management in the 1980s. As technology grew into the 1990s, the Internet became widely available through dial-up means. Some project management entities created.
7. 2000 to Present: Rise of Automation and Maturity of Efficiency	As computer-controlled options and complex algorithms were developed, project managers began to complete more work in less time with fewer errors than ever before. The growth of the Internet led to web-based project management applications being developed. Today, web-based project management applications can be seen on mobile devices, individual computers, and wide-scale ERP systems. There are many factors that must work cohesively to achieve successful results in project management. We can read more about some of the key elements of project management success in this e-Book.

Timeline Facts to Remember

1. Project management has been around since ancient times.
2. The Transcontinental Railroad was the first true project management undertaking in modern times.
3. The Industrial Revolution and associated labour changes drove the need for better project management tactics.
4. Henry Gantt used charts to monitor and manage projects.
5. Frederic Taylor authored a publication on better project management tactics to improve the efficiency of skilled labourers. Many successful organizations still use these tactics today.
6. PERT and CPM were two of the first mathematical formulas for discerning project risk.
7. Computers gave project managers new ways to compute risks and manage projects.
8. The Internet gave birth to mass communication and efficiency across all projects and in greater detail to the devices of today.

Networks (Program Evaluation and Review Technique (PERT))

In project management, the Project Evaluation Review Technique, or PERT, is used to identify the time it takes to finish a particular task or activity.

The PERT chart is a graphical representation of timeline that displays all of the individual tasks necessary to make closing of the project. This is most preferred project management tool; the PERT chart is often preferred to the

Gantt chart because it identifies task dependency.

Origin

Originally developed in 1958 for use by the U.S. Navy Special Projects Office, the PERT technique is a project management system designed to assist in planning large and complex projects. Starting with an overall program evaluation, it promotes in-depth analysis of any project before the kick-off date.

The Program Evaluation Review Technique (PERT) is a statistical tool which was first developed in the United States in the year 1958. This method is adopted to analyses the tasks involved in the project by breaking down the individual tasks which helps in the analysis.

The PERT chart is a useful tool as it allows the project managers to evaluate the time and resources of project required for the project completion.

The PERT can be implemented in three steps.

a) The first step is to find out the tasks required and the order in which these tasks are performed.
b) Next step is to initiate a networking diagram which forms the chart.
c) The last step is to calculate the course and the slacks in time while completing the project.

Program Evaluation Review Technique (PERT) charts can be interpreted by the direction of arrows which indicates the flow and sequence of events required for completing the given project. There are dotted lines on the chart which indicate dummy activities. All in all, the PERT chart determines the time it requires to finalize the project while also helping to analyses the various tasks that form the project.

PERT Chart

A program evaluation chart is a graphical representation of timeline of project that displays all of the individual tasks necessary to end the project.

As a project management tool, the PERT chart is often preferred to the Gantt chart.

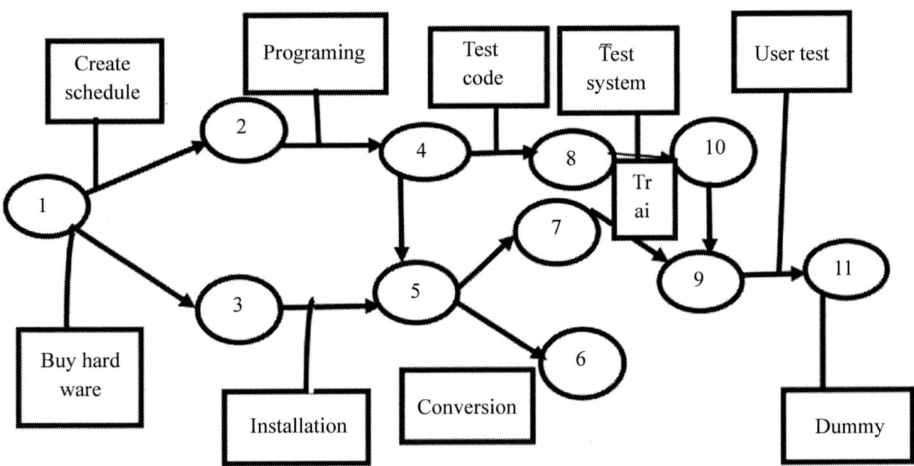

Numbered rectangles are nodes represent events or milestone.

Directional arrows represent dependent task that must be completed sequentially.

Diverging arrows direction (1-2, 2-3) indicates possible concurrent events.

Dotted lines indicate dependent tasks that do not require resources.

Critical Path Method (CPM)

The CPM is a strategy to make survey to find out whether project has the quality of adaptability in situation within required time covering all tasks.

Key Elements

1. Earliest Start Time (ES)	The is initial stage in the project. Activity will start at this stage. It is not possible to decide without knowing the dependencies.
2. Latest Start Time (LS)	It is last point before start. Care is taken to stick to time without deviation.
3. Earliest Finish Time (EF)	Completing of project as per earliest start time.
4. Latest Finish Time (LF)	The end point is calculated using its duration and latest start time.
5. Float	The word "float" points out duration of activity can be prolonged or postponed without affecting its task.

Usefulness of the Method

The approach can indicate priorities, distributing resources, and scheduling projects.

The Reasons:

Improves Future Programming and Planning

Future planning may be planned from experience and knowledge CPM. This will help to draw future plan correctly.

Facilitates to handle resources more carefully and optimally.

It lets project managers prioritise tasks, giving them a better understanding of how and where to deploy resources.

Helps Avoid constraints

Constraints in implementation of project can cost precious time The method CMP has advantages to eliminate constraints.

CPM can help to determine the following:

1. Tasks required for completion
2. Dependencies between tasks
3. The duration to complete an activity

Our approach is now to know two concepts i.e. Event and Activity. We can have a look at the network diagram (which is also the output) of the process.

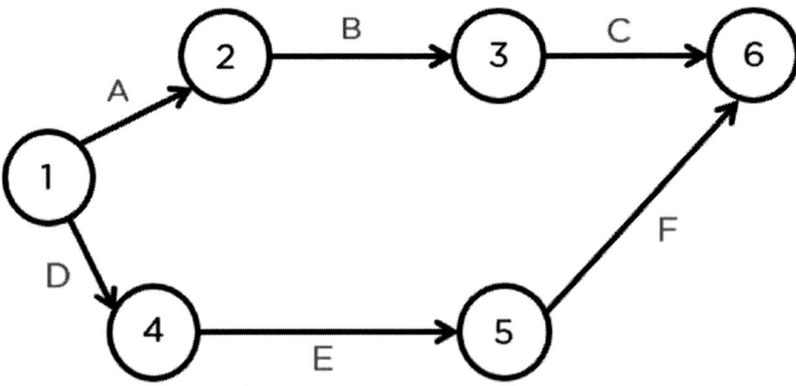

Events and Activities and their significance.

Event

Circles are. Event 1 is the tail event and Event 2 is the head event. In the case of our example, the events are 1, 2, 3, 4, 5, and 6. The nodes 1 and 2, and the connection between them, 1 will be referred to as the tail event, and 2 will be referred to as the head event.

Similarly, for 2 and 3, 2 is the tail event, and 3 is the head event.

Activity

Activities are a resources like time, money, and energy required to complete the project. In the case of our example, A, B, C, D, E, and F represent the activities taking place between their respective events.

Dummy Activity

A dummy activity represents a relationship between two events. In the case of the example below us, the dotted line represents a relationship between nodes 4 and 3. The activity between these nodes will not have any value.

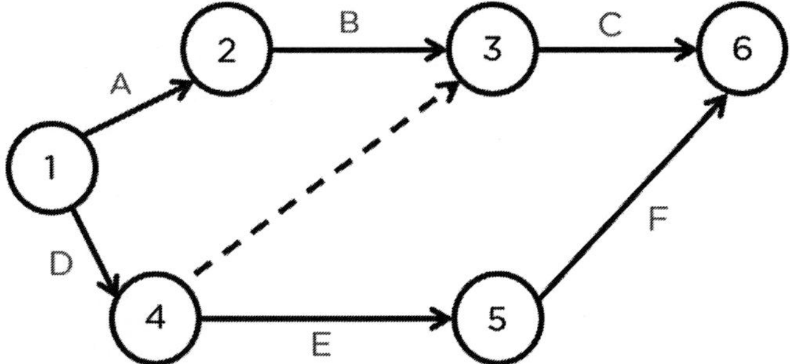

Other Rules to Consider

1. The network should have a unique starting and ending node. In the case of our example, event 1 represents a unique starting point and 6 represents the unique completion node.
2. No activity can be represented by more than a single arc (the line with an arrow connecting the events) in the network.
3. No two activities can have the same starting and ending node.

Now, let's talk about the process of the Critical Path Method with an example.

The Critical Path Method

The objective of the question below is to determine the critical path, based on the information available, like activity, immediate predecessor, and duration (which in this case, we'll take as months).

Activity	Immediate Predecessors	Duration (Months)
A	-	3
B	-	4
C	-	6
D	B	3
E	A	9
F	A	1

G	B	4
H	C, D	5
I	C, D	4
J	E	3
K	F, G, H	6
L	F, G, H	3
M	I	6
N	J, K	9

First, let's analyse the activities and their immediate predecessors.

Activities A, B, and C don't have any immediate predecessors. This means that each of them will have individual arcs connecting to them. First, we'll draw nodes 1 (which is the starting point) and 2. We'll add the activity on the arc, along with the duration.

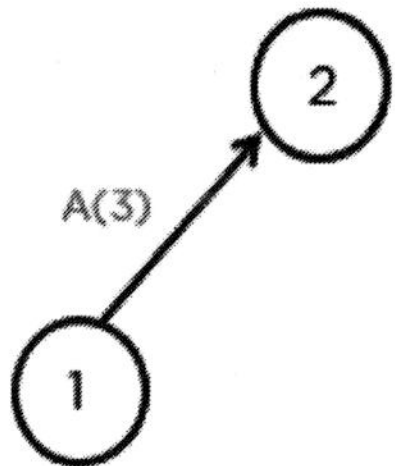

We'll have to also keep in mind that A acts as the immediate predecessor for both nodes E and F. Similarly, let's draw the arcs for nodes B and C.

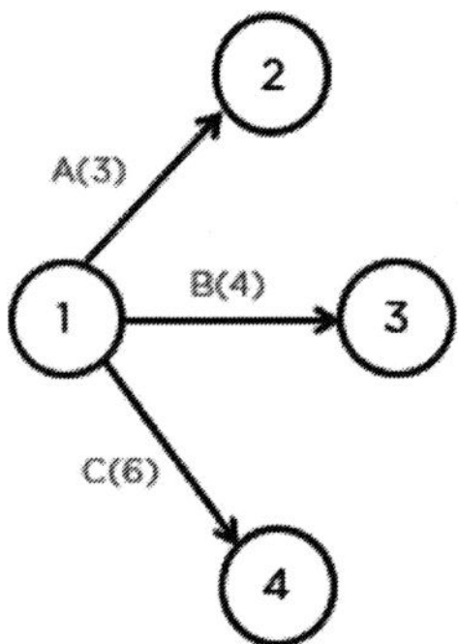

Before we can draw the nodes for activity D, a quick look at the table will tell us that it is preceded by activity B and that a combination of activities C and D act as immediate predecessors for activities H and J. This means that both activities C and D have to connect at some point. That's why we'll be drawing an arc from events 3 and 4.

The pros and cons of the critical path method (CPM) are as follows:

Effective Communication: All phases of a project's life span must be considered when creating critical path method schedules. The program's structure becomes more achievable and firmer when the skills shared by various team members are integrated.

Easier to Prioritise Tasks: Project managers can more effectively prioritize tasks and estimate the float of each one by determining the critical path. Float indicates the amount of time a task may be put off before it affects when it will be completed. A lower float indicates a greater priority.

Accurate Scheduling: CPM is a popular and dependable methodology for enhancing the precision of project schedules. Several project managers utilize CPM with the Programme Evaluation and Review Technique (PERT), which supports teams in estimating overall project length.

Better Visualisation: Gantt charts and CPM network diagrams, which show critical path timelines, can help project managers understand a project's timeline and progress more quickly. These visual tools allow them to understand a project's direction more intuitively than a less eye-catching alternative.

Cons of Using CPM in Project Management

Some of the cons of using the critical path method are as follows:

Multiple Complexities: Several moving elements and detailed computations are involved in the CPM. The software may automate the computations, but entering accurate data requires thorough research and leaves room for human error.

Limited Applicability: Not every project type is suited to the critical path method. Projects requiring creativity, like product design or research work that tend to come along in unforeseen forms fail to lend themselves well to CPM.

Less Understanding of Resources: The critical path method also lacks sufficient understanding of how resource limitations impact project schedules. The accessibility of equipment or labour resources is not considered in the network diagram or CPM schedule.

Step by Step Guide on How to Find the Critical Path Examples

The time of essential and non-critical tasks can be used to identify the critical path. This is a list of the steps, along with examples.

Sl. No	Steps	Work to be Done
1	List Activities	List all the venture exercises or errands essential to create the expectations utilizing a work breakdown structure. The rest of the CPM is built on the list of activities provided in the work breakdown structure
2	Identify Dependencies	Choose the occupations based on our work breakdown structure that is interconnected. This can identify any task that can be finished alongside other chores. These task dependencies are based on the example mentioned above: i) Task A is necessary for Task B. ii) Task B is necessary for Task C. iii) C and D can be carried out concurrently. iv) Task D is necessary for Task E. v) Task F depends on Tasks C, D, and E. An activity sequence, which will be utilized to identify the critical path, is a set of dependent tasks.
3	Create a Network Diagram	The network diagram, a flowchart that depicts the order of tasks, must then be created from the work breakdown structure. Each task should be represented by a box, with task relationships shown by arrows. Until the overall project timetable is determined, we will add other time-bound components to the network diagram.
4	Estimate Task Duration	We must first estimate each activity's length before determining the critical path, which is the longest series of critical tasks. Try these to get a sense of the time: Using knowledge and experience to make educated assumptions Estimating using information from past projects Estimating using established industry practices
5	Forward pass	i) This method uses a previously stated start date to determine early start (ES) and early end (EF) dates. ES is the direct ancestor with the highest EF value; EF is calculated as ES plus duration. The calculation begins at ES of the first action with 0 and moves forward through the schedule. Establishing ES and EF dates enables early resource allocation for the project. ii) A backward pass determines the dates for late starts (LS) and late finishes (LF). LF is the lowest LS value among immediate successors, and LS is LF duration. The calculation begins with the final action on the timetable and works its way backward. The scheduling flexibility of each activity can then be determined using the early and late start and end dates
6	Calculate the Critical Path	Although the critical path can be determined manually, adopting a critical path algorithm can save time.

PERT and CPM: Differences

The PERT and CPM are different. PERT is a management technique that explains planning, scheduling, organizing, coordinating and controlling but CPM is statistical technique of project management well defining planning, scheduling, organizing, coordinating and control of activities.

Difference Between PERT and CPM

PERT

PERT is appropriate technique which does not specify the time for scheduling, organization and integration of different tasks within a project. It provides the blueprint of project and is efficient technique for project evaluation.

CPM

It is a technique indicates time required for completion of project is already known. It is majorly used for determining the approximate time within which a project can be completed. It is the largest path in project management which always provide minimum time taken for completion of project.

PERT and CPM: Differences

PERT	CPM
1. PERT is used to manage uncertain (i.e., time is not known) activities of any project.	1. CPM is that technique of project management which is used to manage only certain (i.e., time is known) activities of any project.
2. It is event-oriented technique which means that network is constructed on the basis of event.	2. It is activity-oriented technique which means that network is constructed on the basis of activities.
3. It is a probability model.	3. It is a deterministic model.
4. It majorly focuses on time as meeting time target or estimation of percent completion is more important.	4. It majorly focuses on Time-cost Trade off as minimizing cost is more important.
5. It is appropriate for high precision time estimation.	5. It is appropriate for reasonable time estimation.
6. It has non-repetitive nature of job.	6. It has repetitive nature of job.
7. There is no chance of crashing as there is no certainty of time.	7. There may be crashing because of certain time foundation.
8. It doesn't use any dummy activities.	8. It uses dummy activities for representing sequence of activities.
9. It is suitable for projects which required research and development.	9. It is suitable for construction project

V. Advantages and Disadvantages of PERT

Advantages	Disadvantages
1. PERT helps in planning for big projects The project manager may easily schedule project activities with the aid of the PERT system. This method is used more frequently in complex, large-scale project work.	1. Time-oriented method PERT is a time-focused methodology, so completing tasks or projects by the deadline is crucial. If it doesn't, an issue might develop.
2. Critical path visibility in PERT The critical path will be clearly displayed using the PERT approach. The crucial path is the path that contains tasks that cannot, under any circumstances, be put off. The project manager will be able to make quick, effective decisions that will improve the performance of the project with the aid of proper knowledge of the stack values and minimal dependency circumstances.	2. Subjective analysis In this case, the project tasks are identified in accordance with the information at hand. It is challenging in PERT projects since it applies to the lone new project area that is not recurring in nature, making the information collection subjective in nature.
3. Analysis of activity in PERT The PERT networks' activity and events are examined. Both of these are examined separately and jointly. This will paint a picture of the project's likely completion and the budget.	3. Inaccuracy in prediction Since PERT lacks historical data for a project's framework, prediction is necessary. In the event that the prediction is incorrect, the project could be ruined.
4. Coordination of PERT The several construction organization departments will supply the data for the PERT operations. The departments are well integrated, which will aid in enhancing the project team's capacity for planning and decision-making. The management of the project's activities will be improved by combining quantitative and qualitative values from a lot of data. Additionally, this will enhance communication within the organization's various departments.	4. Costly process In terms of the amount of time, study, forecast, and resources used, it is too expensive.
5. The what-if analysis By appropriately assessing the critical path, it is possible to learn about the potential outcomes and the various levels of uncertainty from the project activities. Permutation and combination operations are carried out for these different sets. The most advantageous mix of them is taken into account. The set that is selected will have the lowest cost, most economical operation, and best outcome. The risk connected to any activity can be determined with the use of this analysis.	5. Labour intensive PERT analysis is labor-intensive in nature. Large and complex networks form as a result of the possibility of a rise in project activities and the emergence of various task dependencies. This method won't work well for the project if two tasks share resources.

Advantages and Disadvantages of CPM

Advantages	**Disadvantages**
1. Better Communication Critical route method schedules call for feedback from important parties at every stage of the project's lifespan. The timeline becomes more realistic and solid from the outset when the skills of diverse team members & subcontractors, such as architects, electricians, and construction managers, are combined.	1. Increased Complexity CPM drawbacks include increased complexity and using a method with numerous moving elements called the critical path. Although the computations can be automated using software, accurate data entry involves extensive investigation and still carries the danger of human error.
2. Ease in Prioritization Prioritization is made simpler since project managers may more easily define priorities and calculate the float of each work by identifying the critical path. Slack or float measures how much time a task may be put off before it affects when it will be finished. The float of non-critical activities is positive, whereas the float of critical path tasks is zero. Teams can determine the priority of each work by calculating its float. The priority increases as the float decreases.	2. Reduced Applicability Not all projects can benefit from the critical route approach. For instance, CPM demands repeatable and predictable timelines. For creative projects that frequently come together in unforeseen ways, such as product designing or research work, CPM is not a suitable fit. On the other hand, repetitive or autonomous tasks are not ideal candidates for CPM.
3. Improved Scheduling Accuracy The critical path approach is a well-liked and dependable tool for enhancing project schedule accuracy. The PERT analysis helps teams in estimating overall project time and is often used alongside CPM. PERT considers unpredictable events, while CPM concentrates on predictable activities, resulting in three potential timelines: the most optimistic, the most pessimistic, and the most realistic. Project managers can produce the most precise forecasts by combining PERT and CPM.	3. Limited Attention to High-Float Activities When using the CPM, project managers concentrate on the tasks that are on the critical path. Even though the critical route does affect the overall project duration, employing this strategy can make it simpler to disregard low-priority or float activities, which can cause delays. A new building's electrical system installation, for instance, is not a critical path task because it may be completed over a long period of time. However, project managers can still affect the completion schedule if they overlook wiring work or put it off for too long.
4. Better Risk Detection Due to critical route schedules clearly showing the dependencies between jobs, project managers are better able to anticipate the ripple effects of a delay. Compared to alternative systems that do not manage dependencies, CPM offers earlier opportunities for corrections and prevents more surprises.	

5. Greater Flexibility Project managers have the capabilities to quickly adjust the timetable when things don't go as planned, thanks to CPM network diagrams. To compare results and choose the best alternative, project managers might use software that can even simulate the consequences of various adjustments.	
6. More Visual Impact Project managers may quickly comprehend the timing and status of a project by using Gantt charts and CPM network diagrams to illustrate critical path schedules. Team members and project managers can have a more intuitive sense of a project's trajectory using these visual tools than they might be able to use a less visually stimulating option.	

Networks Terminology-Activity, Dummy Activity

A Dummy activity is a type of operation in a project network which neither requires any time nor any resource. It is an imaginary activity shown in a project network to identify the dependence among operation. The representation and more features of dummy activity in network analysis are explained below.

A dummy activity can be either critical or non-critical. It becomes a critical activity when the earlier start time and the latest finish time are equal.

Features of Dummy Activity

Dummy activity forms a connecting link for control purpose or for maintaining the uniqueness of the activity. A dummy activity is used to maintain the network logic and to avoid ambiguity.

An activity followed by a dummy activity can only be completed once the activity or a dummy activity can be either critical or non-critical. It becomes a critical activity when the earlier start time and the latest finish time are equal activities preceding the dummy activity is completed.

Representation of Dummy Activity

A dummy activity, being an imaginary or redundant activity, is represented by a dotted line with an arrowhead. The numbers on the terminal nodes are used to represent a dummy activity. A dummy activity is clearly explained by an example in Figure 1.

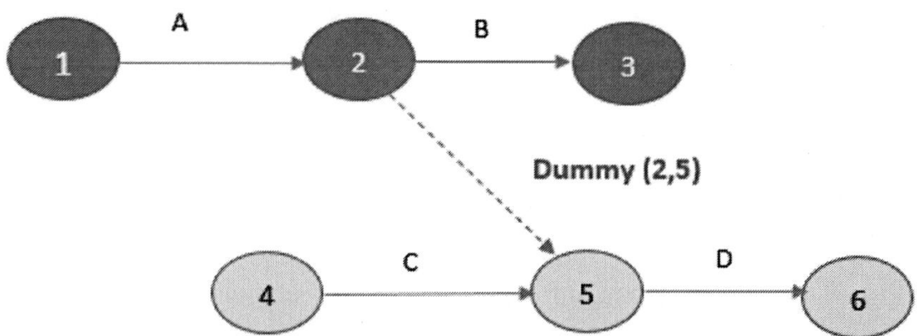

Uses of Dummy Activity

The dummy activity serves the following purposes in a project network:

1. Grammatical Purpose
2. Logical Purpose

Grammatical Purpose

A dummy activity can be used to prevent two arrows with a common beginning and end points. This can be explained by an example.

Consider the arrows of activities A and B. Both start from node 1 and end at node 2. This arrangement is difficult to conduct computations and the network loses its uniqueness in its identification. Such inconvenience causes frequent mistakes during network analysis.

Logical Purpose

It is difficult to represent an activity having two sets of operation running parallel to each other in a network. The use of a dummy activity helps to give a logical representation without difficulty in interpretation.

Rules for Using Dummy Activity in Network Analysis

Too much use of dummy activities in a network creates confusion. Initially, liberal use of dummy activities is followed to fulfil the requirements of inter-relationships between activities. This network is later modified stage-by-stage by removing unnecessary dummies to finally obtain a simple network. This is explained by an example below.

Earliest Start Time (EST), Latest Start Time (LST), Critical Path Terms Used

1. Earliest Start Time (EST)	The earliest start time (EST) is the earliest possible time to start activity
1. Latest Start Time (LST	The last date to start the task without reason for delay
3. Critical Path	The longest path (in time) from Start to finish

Critical Activity, Optimistic time (To), Pessimistic time (Po),

Critical Activity

A critical activity in project management is any task that stakeholders must complete on schedule to meet project deadlines. Delays in one activity will affect subsequent tasks.

How to Find Critical Activities and Tasks in CPM

A critical path in project management is the longest sequence of dependent tasks that stakeholders must in order to deliver a project on schedule. By identifying the critical path, we can determine the earliest date for finishing project.

Critical activity in project management is any task that stakeholders must complete on schedule to meet project deadlines. Delays in one activity will affect subsequent tasks.

The terms critical activity and critical task are often used interchangeably in project management. The two types of critical activities are :

Dependent Tasks: This category includes any activities we must complete in a certain order.

Parallel Tasks: We can run these activities concurrently with other tasks.

Importance

The importance of Critical path activities are to be taken seriously because they can throw the whole project off schedule. The key distinction between critical vs. non-critical activities is whether a late start or finish will push back the project completion date.

Related Concepts

1. Earliest Start Date	Identify this date based on time it would take to finish the dependent tasks immediately preceding
2. Earliest Finish Date	Assess the minimum time required to complete the task, and calculate this date based on start date
3. Latest Start Date	Determine the latest date to begin without jeopardizing the schedule
4. Latest Finish Date	Latest date we can complete a task.

Project managers assign a float (also called to as slack), which is can be delayed without impacting the project schedule with zero float.

CPM works best when there's a high level of certainty about how long activities will take to complete. When project managers aren't sure of time factor of some tasks, they'll often use CPM in with the project analysis.

In PERT, we calculate three durations: the optimistic duration, likely duration, and the pessimistic duration. Using a weighted average of the three durations, PERT creates an estimated duration.

Divide the Project into Tasks: Break down the project into basic activities, keeping a high-level focus. From there, break down activities into more detailed subtasks, then use a 'Work Breakdown Structure' (WBS) to arrange activities further into manageable pieces of work.

Establish Dependencies: To establish dependencies among tasks, ask ourself the following questions for each task:

i) What tasks need to be finished before this task begins?

ii) What tasks can be finished at the same time as this task?

iii) What task should begin immediately after this task is completed?

Create the Critical Path Analysis Chart (CPA):

The CPA chart, also known as the network diagram, visually represents activities and their dependencies. Once we have established activities and dependencies, we can create our critical path analysis chart, either by drawing it by hand or using software.

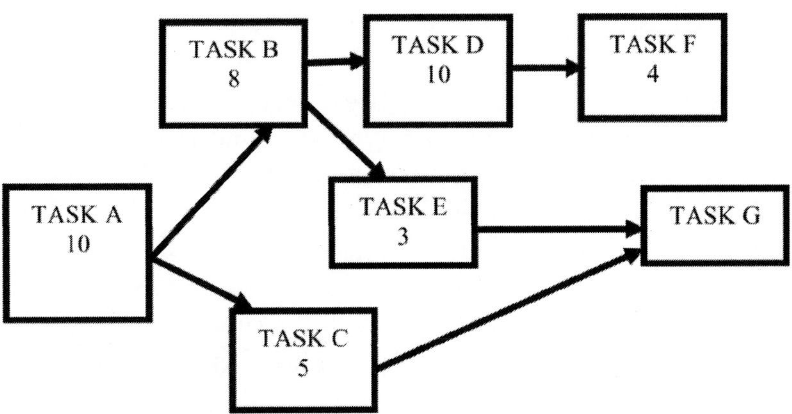

Completion Time for Each Activity: Estimate the time we believe each activity will require. Project managers often use scheduling software to set a realistic timeframe for task duration, and to identify places to reduce the duration of critical activities in CPM.

To create a realistic schedule, many project managers estimate completion time using the following three-point system:

i) Best-case scenario, or shortest time frame
ii) Most likely scenario, or most realistic time frame
iii) Worst-case scenario, or longest time frame

Identify the Critical Path: There are two ways to identify the critical path:

i) Use our network diagram and draw the possible longest path.
ii) Use the forward pass/backward pass method, in which we identify the earliest start, finish times, and latest start and finish times.

Update the Diagram Critical Path throughout the Project: As the team completes activities over the course of the project, update diagram to show the actual time for completion rather than our projections. Doing so will help to create a more realistic schedule as the project progresses and, possibly, identify a new critical path.

Optimistic time (To): "The least amount of time it can take to complete a task. Pessimistic time. The maximum amount of time it should take to complete a task. Most likely time. Assuming there are no problems, the best or most reasonable estimate of how long it should take to complete a task".

Pessimistic Time (P): "The maximum amount of time it should take to complete a task. Most likely time. Assuming there are no problems, the best or most reasonable estimate of how long it should take to complete a task".

Most likely time (TM), Expected time (TE), Float or Slack, Event Slack

Most likely time (TM): "Most likely time, assuming there are no problems, the best or most reasonable estimate of how long it should take to complete a task. Expected time-Assuming there are problems, the best estimate of how much time will be required to complete a task".

Expected time (TE): "The expected time (te) is computed using the **formula (o + 4m + p) ÷ 6**. Once this step is complete, one can draw a Gantt chart or a network diagram. A Gantt chart created using Microsoft Project (MSP)".

Float or Slack: "Slack, also known as total float, is the amount of time that an activity can be delayed without delaying the project end date or an imposed milestone. Float, also known as free float, is the amount of time that an activity can be delayed without delaying the early start of its successor activity."

Event Slack: "The head event slack of an activity in a network is the slack at the head (or terminal point) of an activity. In other words, head event slack of an activity in a network is the difference between the latest event time and earliest event time at its head (or terminal point or node)".

Crashing Critical Path, Acclivity Table

Clarification: The critical path is the shortest time a team can finish a project. Crashing involves reducing this time by adjusting tasks that directly impact a project's duration, called critical tasks. Here are common knowing the answer to, "What is crashing in project management?" can help to understand its purpose and apply it effectively in our role. Crashing in project management means considering how a new work plan may influence project costs. Its purpose is typically to obtain a reduced work schedule for the least incremental cost. Successfully crashing a project involves estimating the possible return on investment to ensure the decision can help to reach a project's goals.

To reduce a project's duration, it's often necessary to determine its critical path and tasks. The critical path is the shortest time a team can complete a project. Crashing involves reducing this time by adjusting tasks that directly impact a project's duration, called critical tasks.

Here are common ways a project team may achieve this reduced timeline: adjusting tasks that directly impact a project's duration, called critical tasks.

To reduce a project's duration, it's often necessary to determine its critical path and tasks. Here are common ways a project team may achieve this reduced timeline:

1. Allocating additional resources to critical tasks
2. Motivating project members using incentives
3. Outsourcing project tasks

Reasons of Crashing a Project

1. To prevent project delays.
2. To receive bonuses for completing the project early.
3. To handle multiple projects effectively.
4. To use resources productively: Crashing projects can help you use idle resources well.
5. To train a project member.

How to Crash a Project

If we are considering project crashing, we can follow these steps to ensure to successfully reduce the schedule:

Create and reflect (critical path)

Start by creating a network diagram of the critical path, which visually represents a project's schedule and each task's duration. We can include the earliest time to start and complete a task. Alternatively, we may show the latest start and completion time. Doing these can help we identify how long we may crash each activity. Reflecting on the critical path involves identifying tasks we can complete quickly, especially if additional resources become available. We can expect to apply your analytical and critical thinking skills to complete this step.

Evaluate the benefits and costs involved

Consider the impact of reducing each task's duration. For example, if hiring a project member can help reduce a project's delivery time, consider the training time and the additional cost of implementing this strategy.

Determine possible project solutions

Next, identify possible schedules that may help to complete the project early. For example, we may reduce the duration of all critical tasks or only those that take time. Determining cost-effective solutions typically requires us to start with the task that has the least crashing cost. It also involves evaluating the possible risks of implementing the new project schedule.

Create a budget and select the most cost-effective solution

A budget can help to choose a new schedule from the possible plans. For example, if we have six potential solutions, we may choose the strategy that offers more benefits for reduced costs. It's also important to update the project's delivery timeline and create a new diagram of its critical path.

Implement the new project schedule

Continue with the new timeline by informing team members about it. This can also help to review the strategy and ensure our planning steps are accurate. Upon receiving approval from upper management, we can start planning for the additional resources you may require.

Activity Table

Clarification: "Most likely time A PERT chart is made up of nodes and directional arrows. Nodes are numbered boxes or circles. They represent an event or milestone in the project. The arrows are the tasks or activities that must be done before moving on to the next event or milestone"

Activity table for example

Role	Activity	Environment	Activity Completion	Property/ Notifica-tion	Act	Act competition	Activity structure	Time
Student	Discuss the case (step 1)	Synchronous Conferencing service Study land scape	User choice	-	Act 1	When all students are done	-	-
Student	Discuss additional Information (step 2)	Synchronous Conferencing service Study land scape	User choice	-	Act 2	-	-	-
Student	Combine results of both discussion (step3)	Synchronous Conferencing service	User choice		Act3	When chairperson is done	--	-
Chairperson	State problem	Synchronous Conferencing service	User choice	Upload file	ACT3	-	--	-
Student	Formulate casual explanation (step 4)	Synchronous Conferencing service Study land scape	User choice	--	ACT 4	When all students are done	--	-
Student	Make differential diagnosis (Step 5)	Synchronous Conferencing service	User choice		ACT 5	When all students are done	-	-

Student	Discuss how to get more certain diagno-sis (STEP 6)	Synchronous Conferencing service	User choice		Act 6	When all students are done		
Student	Develop therapy (step 7)	Synchronous Conferencing service Study land scape	User choice		ACT 7	When chairperson is done		
Chair Person	State therapy	Synchronous Conferencing service Study land scape	User choice	Upload file	ACT 7	-	-	-

Danglers, Normal Time

Dangling activities (also known as dangles) are loosely-tied activities in project schedules. They are activities with either open start dates or open-end dates. All activities, except the first activity of a network, need to have a predecessor; otherwise, they will have open start dates.

Dangling activities (also known as dangles) are loosely-tied activities in project schedules. They are activities with either open start dates or open-end dates. All activities, except the first activity of a network, need to have a predecessor; otherwise, they will have open start dates. Similarly, all activities, except the last activity of a network, need to have a successor; otherwise, they will have open end dates (also known as open-ended or open-end activities).

As noted above, every project activity and milestone except the first and last ones must have at least one predecessor and one successor. The first activity of a network is the notice to proceed milestone and an example of the last activity of a network is the milestone represents the project finishing date. It is recommended that any project schedule starts with a start milestone and finishes with a finish milestone to ensure proper logical ties can be built into the network.

A project schedule that contains dangling activities has deficiencies because its logic is incomplete. This flaw makes the schedule unreliable and inaccurate because the schedule has not fully developed and some activity dependencies (i.e., logical ties) have not been properly identified. The four main types of activity ties include finish-to-finish (FF), finish-to-start (FS), start-to-start (SS) and start-to-finish (SF).

1. With Dangling activities managing project delays become difficult as identification of these delays are a challenge in the absence of estimated start and end dates.
2. The dangling activities make the calculation of float/slack become challenging as we cannot identify the early or late start dates of connected activities.
3. Logic Driven project schedules can suffer from two kinds of open-ended or Dangling Logic, which makes the resulting schedule unreliable for dates or float analysis.

Normal Time

Normal time is initially planned to finish the activity. Crash time is the time that activity will take if the additional resources are expended. Crash time is NOT the time saved, rather it is the reduced time applied. Crash time will always be a number smaller than normal time.

Calculate normal project duration

When we build a schedule, we need to understand how to estimate duration. If everyone worked eight hours per day, and was 100% productive for all eight hours, we could easily calculate our estimated project duration by taking the number of effort hours, divided by the number of resources.

Steps in Network (Rule of Preparation)

Steps to calculate estimated duration

Estimate the productive hours per day: Normally the first step is to determine how many productive hours of work we can count on each person working per day over time. Using a factor of 6.5productive hours take into account socializing, ramp-up time, going to the bathroom etc.

Determine how many resources will be applied to each activity: In general, the more resources we can apply to activities, the quicker the jobs can be completed. Obviously two resources may enable to complete an activity faster than one person, but twice as fast. Similarly, a third person may allow the task to be completed sooner, but not in one-third the time. However, at some point, adding resources will not make the activity complete any sooner, and in fact, may make it go longer.

Factor in available workdays: Take into account holidays, vacations and training. The was not in the productivity factor in the first item, since this non-project time can be scheduled and accounted for in advance. For instance, on a three-month project, one team member may be out for two vacation days, while another may also have ten days of vacation. To make our schedule more accurate, take into account any days that we know our team will not be available to work on the project.

Calculate delays and lag-times: Some activities have a small number of effort hours, but a long duration. For instance, a deliverable approval may take one hour, but might take two weeks to schedule the meeting.

Identify resource constraints: When we build our initial schedule, we identify the activities that to be done sequentially and those that be done in parallel. If we have enough resources, all of the parallel activities can, in fact, be done in parallel. However, we can only do the activities in parallel if we have the right resources available at the right time. The activities that can be done in parallel; however, they need to be worked on sequentially because only one person has the right skills to do the work – even though other resources are available.

Document all assumptions: Therefore, it is important to document all the assumptions we are making along with the estimate.

7

Program Evaluation Tools Bennett's Hierarchy of Evaluation

Bennett's Hierarchy

To set the scene for this we head back to the 1970s when Claude Bennett worked as a specialist in evaluation with the extension service. He wanted to make sure that extension was collecting information that was able to illustrate the effectiveness of extension, and help extension improve its services to clients. So, he developed seven categories of criteria for evaluating extension programmes based on what he called a 'seven link chain of events. He put these into a hierarchy literally in his article describing this-as a diagram of seven steps.

Background and Description; Relation Between Programme Objectives & Outcomes at 7 levels of Bennett's Hierarchy

The Hierarchy describes a series of staircase levels of evidence of program impacts, beginning at the bottom step with "inputs" (allocation of resources to a program) and progressing to the top step, "end result" (measuring impacts of a program on long-term goals or conditions).

Evaluation shouldn't be just the add-on at the end of a project, it's a core skill for enablers of change like ourselves. So, let's explore one framework for evaluation that's been around for some time-Bennett's Hierarchy.

Programme and Objectives-Interrelationship

Program objectives explain the potential applications of the knowledge and skills acquired in the program; seek to help students connect learning across various contexts; situate the particular program in the context of the discipline as a whole; and are often broader in scope than the program-level learning outcomes.

Program objectives specify how program goals will be achieved and should include a method for evaluating results. While program goals should clearly state the intentions of a program, objectives should describe the mechanisms and strategies used to accomplish those intentions.

Differences Between Objectives and Outcomes

Objectives are intended results or consequences of instruction, curricula, programs, or activities. Outcomes are achieved results or consequences of what was learned.

However, an objective is an intermediate result that must be realized to reach a goal. The definition of an objective is usually more focused than that of a goal and is typically more subject to being measured. Objectives can be further assessed and evaluated using performance measures.

The 7 levels of Bennett's Hierarchy

Bennett's hierarchy contains seven sequential steps (input, activities, participation, reaction, knowledge, skills, opinions and aspirations-KASA (Knowledge, Attitude, Skill Aspiration) practice change, and end results/ social, economic, environmental conditions. SEEC

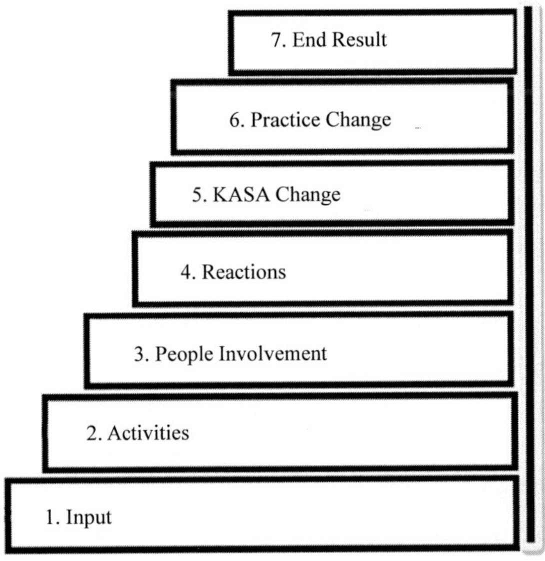

Figure 1. A hierarchy of evidence for Program Evaluation

So, let's go through each of these seven levels.

Starting at the bottom we have level 1 and this is inputs. This is resources that are required to run extension programmes, usually things like time, money and extension personnel.

Next up is level 2 and this is the activities-the extension activities-these could be workshops, field days, or resources produced.

Then for level 3 we have participation. This is a common thing for extension programmes to monitor-we count how many people turn up to our workshops or field days or how many saw our work on Twitter or other social media. This is what people like reporting and it's generally easy to do so!

Level 4 is reactions that is the response of the participants to the activities. It may be that they get more interested in finding out about something, that they learn how to do something-it's the immediate response.

Next, we have level 5 and this is where it starts to get really interesting. This is KASA level change; Knowledge, Attitude, Skills and Aspirational change. This is getting towards the heart of extension we want people to be able to change and these are the precursors to practice change which is the next level.

Level 6 is practice change. This is the sole aim of extension - it's why we do what we do. This is the adoption of a new practice, technology, or skill.

And finally, we get to level 7 which is the end results. This is the big picture outcomes, the social, economic, environmental or cultural change driven by the practice change.

So, we have the seven levels of Bennett's Hierarchy. Z Well Claude had some pretty good tips for thinking about evaluation using the hierarchy. He pointed out that when we're thinking about evaluation of extension, the higher up the steps we can gather evidence, the stronger the evidence of impact. So, the hierarchy forces us to ask questions about what level will we aim to get to

Bennett was also a realist too though, because he also pointed out that the higher in the hierarchy we go, the greater the difficulty and cost of obtaining the evaluation data, especially hard data. So, what's our budget like and what is realistic to try and gather data on?

These questions are often overlooked when it comes to evaluation, so it's good practice as an enabler of change to have the chance to work this through and decide what is possible, and why, when evaluating a change programme. This means we can be clear about the change we can track and the amount of time and effort required.

In fact, Claude went on to use his Hierarchy as a programme development tool and a programme evaluation tool. Essentially one can use the hierarchy as a way of work shopping the ultimate aims of a programme, and work down the levels to figure out what is needed at each level. Then we can use the hierarchy to figure out what data we can gather to determine how well the programme is going. There are few resources that cover this process that we'll put links below.

Bennett's Hierarchy Logic Model

Bennett's Hierarchy is a type of logic model tailored for programs that provide information or education. Most of us are familiar with the Kellogg logic model which has buckets for inputs, activities, outputs, short, medium, and long-term outcomes. Bennett's Hierarchy provides more specialized terminology for educational programs. We know "KASA," that's Bennett's Hierarchy.

Bennett's Hierarchy provides more guidance about what educational programs should be targeting for outcomes. Thus, it is more intuitive for practitioners to use. It helps them better plan for how a program will be successful and better describe its successes.

Kellogg-style logic models. Once we get to the outcomes, all the energy is drained and it feels like we are looking into a big black hole of possibilities. How do we figure it out? Bennett's, model on the other hand is typically, structure to guide our thinking about what makes the program successful. We end up feeling like we came up with something useful and powerful. At its core, Bennett's Hierarchy isn't that different from the Kellogg model. It's just a lot more helpful.

Kellogg Logic Model	Inputs	Activities	Outputs	Short-term Outcomes		Med-term Outcomes	Long-term Outcomes
Bennett's Hierarchy Logic Model	Inputs	Activities	Participation	Reactions	Knowledge Attitude Skills Aspirations	Practice Change	Impact

Bennett's Hierarchy describes seven steps that are necessary to get from input (the first step) to impact (the final step).

Inputs: What we are giving? Same as Kellogg model.

Work Activities: This is a indicates of what work to be done. Same as Kellogg model.

Target group or Participation: This is of two things. First, identification of target audience ? Second, number of people involved. For how long, how often, etc. Scope, frequency, duration, intensity. It is a more tailored version of Kellogg's outputs section.

Positive or Negative Reactions: It reveals immediate feeling towards program. For audience the program is interesting, enjoyable, and informative, etc ? We can also call this the participant's experience. These are a part of what Kellogg would call short-term outcomes.

KASA: KASA is an integral part of part of Bennett's Hierarchy. It determines "practice change" The participants should exhibit prior changes in four areas.

Knowledge, Attitude, Skill and Aspiration: This would also be a short-term outcome in Kellogg's model.

Practice Change: Adoption of learned skill in practical life is the aim of educational goal. They should learn new ways of living and doing things. Thais is how the participant's lea earn. They may differ but learning takes place. Educational programs bring hope and change. . This would be a medium-term outcome in Kellogg's model.

Result or Impact: Final achievement is the impact or end point. What it contributes to man and society is the out –come, this is final output in Kellogg's model.

Bennett's Hierarchy is indicated as a staircase with inputs on the base and impact at the top. It is as we are climbing the staircase revealing our outcomes are stronger and further reaching.

Inputs, Activities, Participation, Reactions

Input: Inputs are any resources used to create goods and services. Examples of inputs include labour (workers' time), fuel, materials, buildings, and equipment.

Activities: In simple terms, project activities are the individual actions one must complete to achieve project success. Each activity will count as a project stage consisting of tasks and subtasks that all contribute to the activity's completion

Participation: An act or instance of participating. The fact of taking part, as in some action or attempt participation in a celebration. At the most basic level, participation means people being involved in decisions that affect their lives. Through participation people can identify opportunities and strategies for action, and build solidarity to effect change.

Reactions: The action or state of taking part in something: as association with others in a relationship (as a partnership) or an enterprise usually on a formal basis with specified rights and obligations.

Participant Reaction measures the initial satisfaction of the participant to the program design and delivery. Information is frequently gathered through questionnaires at the end of the session.

The first level of criteria is "reaction," which measures whether learners find the training engaging, favorable, and relevant to their jobs. This level is most commonly assessed by an after-training survey (often referred to as a "smile sheet") that asks students to rate their experience

KASA Changes, Practice and Behaviour Changes, End Results

Knowledge: The facts, information, and skills acquired through experience or education; the theoretical or practical understanding of a subject.

Attitude: It is the disposition, feeling, position, etc., with regard to a person or thing; tendency or orientation, especially of the mind: a negative attitude; group attitudes. Position or posture of the body appropriate to or expressive of an action, emotion, etc.: a threatening attitude, a relaxed attitude.

Skill: the ability to use one's knowledge effectively and readily in execution or performance. Dexterity or coordination especially in the execution of learned physical tasks learned power of doing something competently: a developed aptitude or ability.

Aspiration: The words ambition and pretension are common synonyms of aspiration. While all three words mean "strong desire for advancement," aspiration implies a striving after something higher than oneself.

Significance of Practice and behaviour changes, end results

Practice and Behavioural Change:

The stages are:

1. Pre-contemplation
2. Contemplation
3. Preparation/Determination
4. Action / Willpower (Changing behaviour)
5. Maintenance (Maintaining the behaviour change)

Stage	Activities
One: Pre-contemplation	People are not thinking seriously about changing and are not interested in any kind of help. They tend to defend and stage is called "denial
Two: Contemplation	The people weigh the pros and cons of modifying their behaviour. May think about the negative and the positives aspects associated with change
Third: Preparation/Determination	Symptom of change. Indication of need to do to change their behaviour. Too often, people skip this stage: they try to move into action and may face problem.

Four: Action/Willpower	People are motivated to change their behaviour. This is the shortest of all the stages. This is a stage when people most depend on their own willpower. Mentally, they review their commitment to themselves and develop plans to move ahead.
Five: Maintenance	Maintenance involves being able to successfully avoid any temptations to return to the bad habit. in advance

End Result

End Result: The end result of an activity or a process is the final result that it produces. The end result is very good and very successful.

Advantages and Disadvantages of Bennett's Hierarchy

Clearly Defined Career Path and Promotion Path

When a business has a hierarchical structure, its employees can more easily ascertain the various chain of command. Having clear advancement opportunities can help attract and retain talented professionals. Promotions also help employees experience increased morale, motivation and productivity.

Department Loyalty

Companies with hierarchical structures divide employees into teams and departments. When employees are part of a department, they tend to grow a sense of team spirit and loyalty. These sentiments can help retain employees and encourage team members to work together to achieve the organization's goals.

Efficient Leadership and Communication

Having a hierarchical structure helps employees understand the various levels of leadership. Employees tend to know who to talk to when providing progress updates or reporting issues. Management also benefits, as they can use their authority to effectively delegate tasks and ensure operational efficiency. Overall, hierarchical structures can lead to better communication and collaboration across teams.

Delegation of Authority

Instead of having one entity hold all the authority, a hierarchical structure delegates power to different employees. This distribution benefits the chief executive officer, as they can focus on making decisions that affect the entire organization. Supervisors at the department level have the necessary authority to manage the daily performance of general employees

Encourages Specialization

The various departments within a hierarchical structure allow employees to become specialized in a particular field. Specialization ensures an organization can offer high-quality products and services that stand out from competitors. For example, a newspaper company could separate employees into roles like news reporters, sports reporters and designers. The news reporters can focus on generating breaking news, while the sports reporters specialize in sports writing. The designers can hone their artistic abilities to create attractive news materials.

Disadvantages of Hierarchical Structure

Although implementing a hierarchical structure has many benefits, this type of organizational technique can also have drawbacks.

Disadvantages to Consider

It can be Costly

Because a hierarchical structure requires multiple departments, companies employing this organizational technique employ several managers and supervisors. Employees in these roles have higher education levels and experience, requiring higher salaries. Companies can prevent these higher salaries from affecting their profits by managing their budgets and hiring only the management staff necessary for efficient operations.

Slower Decision-Making

Hierarchical companies tend to involve their managers and supervisors in decision-making. While this amount of expertise can seem beneficial, it can also result in slower actions. This effect is especially prevalent when a fast decision is necessary to respond to urgent business decisions. For instance, a company might involve several supervisors in the hiring process. Differing opinions might lead to slow hiring decisions that lead qualified candidates to apply elsewhere. A hierarchical organization can prevent this issue by delegating decisions to specific management teams and limiting the amount of input across the organization.

Poor Communication

Dividing employees into different departments results in lack of communication. Professionals might deem it unnecessary to communicate with other teams or lack the necessary resources. An organization with a hierarchical structure might prevent this issue by hosting regular cross-department meetings. Prioritizing communication across the organization promotes effective teamwork and ensures all departments have readily available opportunities to collaborate.

Department Rivalry

The division into teams might also cause department rivalry within a hierarchical organization. Employees might feel like competing with their peers instead of working together toward the company's objectives. These various departments could also make decisions that only benefit themselves rather than the company.

8

Logic Framework Approach (LFA)

Logical Framework Approach (LFA) Background and Description, Variation of LFA

Definition The Logical Framework Approach or **LFA** is a systematic and analytical process for objectives-oriented project planning and management

Background: The Logical Framework Approach (LFA) was developed for USAID in the 1960s. Since then, it has been adopted and adapted by many other international development organisations. Among them was the German agency GTZ, which derived its Goal-Oriented Project Planning (ZOPP) from it.

Description The Logical Framework Approach (LFA) was developed for USAID in the 1960s. Since then, it has been adopted and adapted by many other international development organisations. Among them was the German agency GTZ, which derived its Goal-Oriented Project Planning (ZOPP) from it. LFA is widely used today, although the methodology is often used in a more flexible and more pragmatic manner than in the 1970s and 1980s. Also, many approaches known as "Results-Based Management" (RBM) and "Managing for Development Results" are based on the Logical Framework Approach or are at least closely related to it.

The Logical Framework Approach is a systematic and analytical planning process used for the results-based planning of a project (or programme) and for the associated monitoring and evaluation system. The basic idea of the Logical Framework Approach is to condense the planned project mechanism down into a relatively simple, linear Logic Model, using a documented situation and problem analysis as the point of departure. This forms the basis for planning the monitoring and evaluation system, whereby the project's outputs and effects are recorded by means of quantitative or qualitative indicators. Lastly, the project mechanism and the monitoring and evaluation system are summarised in a standardised table (log frame). The Logical Framework Approach is therefore not per se a method of measuring impact. Instead, it helps with planning projects and evaluating them in a goal- and results-based manner.

Objective: The purpose of LFA is to undertake participatory, objectives-oriented planning that spans the life of project or policy work to build stakeholder team commitment and capacity with a series of workshops.

Variation: The Logical Framework Approach (LFA) is a methodology mainly used for designing, monitoring, and evaluating international development projects. Variations of this tool are known as Goal Oriented Project Planning (GOPP) or Objectives Oriented Project Planning (OOP)

It is a methodology mainly used for designing, monitoring, and evaluating international development projects. It is called as Goal Oriented Project Planning (GOPP) or Objectives Oriented Project Planning (OOPP).

Goal Oriented Project Planning (GOPP) or Objectives Oriented (Activities, Outputs, Purpose and Goal & Columns representing types of information about the events Project Planning (OOPP);

(i) Goal Oriented Project Planning (GOPP)

The G.O.P.P. (Goal-Oriented Project Planning) is a tool of management of project in which interactive workshops involving all stakeholders in a project, together with an external moderator, are held at different points in the project lifecycle.

The GOPP Method

GOPP (Goal-Oriented Project Planning) is an innovative tool for project management in which interactive workshops involving all stakeholders in a project together with an external moderator are held at different points in the project lifecycle. The GOPP method can be freely used: it is not copyrighted or patented. It has been used in a wide variety of situations since the 1980s and has proved to be robust and effective in helping groups defining clear objectives and designing related action plans. It is an excellent method for improving team effectiveness.

GOPP aims to improve the quality of the analysis made by the group of partners in the design phase of a project to make the project more coherent and transparent by clarifying the responsibilities of each partner; to provide trust and self-confidence to project partners so to reduce the risk of lack of commitment or failure during the implementation of the intervention.

It aims to improve the capacity of the group of partners to achieve more results in a limited time. How does it work?

A successful GOPP session needs a skilled and independent workshop facilitator or moderator. He ensures that the discussion is always focused and makes sure that all the participants are involved in reaching decisions on an equal basis.

Ahead of time, he agrees with the client what is blocking the participants from making the desired progress and determines which concrete workshop products can be delivered in the time available. The moderator prepares a planning procedure which will lead the group through all the steps needed to solve the issue or problem that blocks their progress and deliver the products. To assist the group, the moderator also uses visualization techniques.

A GOPP workshop is normally attended by a maximum of 20 participants and it may vary a lot in duration, from 1 to 5 days. However, a 2-days workshop is normally enough for reaching significant results with a group.

Benefit from GOPP

Recent applications GOPP may be useful whenever an analysis and decision-making process is at play within a partnership or an organization. Up to now, the situations in which it has been mostly used are the following:

Projects funded or co-funded by the European Commission; especially in the case where a transnational partnership has to co-operate, the use of an external facilitator may be very useful for overcoming language or cultural problems in meetings that are necessarily short;

Local development processes; GOPP may ensure more transparent results to spontaneous development processes initiated by key-actors at a local level; -changing processes within organizations. The commitment and the participation of stakeholders and workforce may be a success factor in enabling organizations to face changing environment and establish higher development goals;

Social and economic research: new ways of interactive meetings or interviews with stakeholders or clients may be conducted within a GOPP-like environment meetings, conferences, workshops: the use of more interactive group-work sessions for involving conference participants in the discussion (round tables, small groups, thematic workshops, etc.) is increasing now also in large conference or meetings.

The GOPP process is designed to

Define realistic objectives which can lead to sustainable achievements;

Improve communication and cooperation between project staff, beneficiaries, and partners;

Clarify the scope of responsibility of implementation; and

Develop indicators for monitoring and evaluation.

ii) Objectives Oriented (Activities, Outputs, Purpose and Goal)

Depending on the organisation, there may be varied definitions and understanding among these key concepts and it is vital for the organisation to agree on the terms and appreciate the differences.

The **objective-** It is what the project is aiming to achieve.

The **output-** It is what the project actually delivers in terms of process, people (skills), and technology

The **outcome-** It is the business change that is a direct result of the output.

The **benefit-** It is the measure of the advantage gained by the organisation through achieving the outcome.

Objective Oriented Activities

Objectives are the specific steps for achieving a goal. Objectives are usually precise, and measurable statements that explain how the goal will be accomplished. Activities provide further detail on how the objective will be achieved by listing specifically what will be completed in a specific timeframe.

Guidance for writing goals, objectives and activities.

Difference between a goal, an objective, and an activity.

Goals are typically broad general statements that describe what the program plans to accomplish. They establish the overall direction and scope of the program and serve as the foundation for developing program objectives.

Activities provide further detail on how the objective will be achieved by listing specifically what will be completed in a specific timeframe.

Objective Oriented Outputs

The messages and classes identified through the development of the sequence diagrams can serve as input to the automatic generation of the global class diagram of the system.

Objective Oriented Purpose

Object-oriented programming (OOP) is a way of thinking about and organizing code for maximum reusability. With this type of programming, a program comprises objects that can interact with the user, other objects, or other programs. This makes programs more efficient and easier to understand.

Objective Oriented Goal

A goal is an achievable outcome that is generally broad and long-term while an objective defines measurable actions to achieve the overall goal. Find out the real differences between the two to inform our team’s strategy.

When it comes to leading a team, setting goals and objectives helps to achieve desired results. From large business goals to small daily objectives, these methods help set up team apart from the competition.

While both are important, goals and objectives differ when it comes to the specific actions we should take. It's essential to understand how to incorporate both within our project portfolio to accomplish big picture plans.

We'll go over the major differences between a goal vs. objective, techniques for each, and cover how to incorporate them into our daily routine.

Goal vs. Objective

A goal is an achievable outcome that is generally broad and longer term while an objective is shorter term and defines measurable actions to achieve an overall goal, while different, the two terms are often used when working on a project. This is because both are essential to planning and executing a project. Both create measurable steps to reach the desired outcome.

Types of Goal

1. Time-Bound Goals	Time-bound goals are focused on setting timely actions. This means they are driven by deadlines and target dates. They provide a high-level explanation for what your team should be striving toward. To be time-bound, a goal must be connected to a specific timeline. Generally, these are long-term, actionable deadlines that connect to a business plan. This type of goal helps teams execute high priority, time sensitive actions.
1. Outcome-Oriented Goals	Outcome-oriented goals are focused on the end result. Rather than focusing on specific deadlines, outcome-oriented goals look to accomplish the action above all else. These goals may result in deadlines being pushed back.
2. Process-Oriented Goals	Achievement focused internal system and processes is preferred over specific outcome-oriented goal. The option is to increase efficiency. so that process-oriented goals can prevail.

Types of Objectives

1. Strategic Objectives	It is an attempt to have purpose-based strategy for present entire vision of the work activities. It is meant for expansion of business. The members move with very clear direction.
2. Tactical Objectives	These are set for short term output. The decisions are taken for both short-term keeping end point at view which is of long term in nature.
3. Operational Objectives	These objectives are action oriented and view result or output. It is checked on daily, weekly or fortnight basis.

Columns Representing Types of Information about the Events Project Planning (OOPP)

1	Task name
2	Project Management
3	Star of Project
4	Definition
5	Analyze requirements
6	Conduct feasibility study
7	Preliminary project plan and proposal
8	Project plan complete
9	Analysis
10	Prepare functional Specialization document
11	Functional specification review
12	Functional specification complete
13	Revised project plan
14	Design
15	Prepare design specification document
16	Design review
17	Revised documents
18	Set up development environment
19	Design process complete

Narrative Description, Objectively Verifiable Indicators

The Objectively Verifiable Indicators show the important characteristics of the objectives and the performance standard expected to be reached in terms of quantity, quality, time frame and location.

Indicators are used to measure activities, outputs, outcome and impact. Indicators' place in the project log frame varies according to the model used by the donor. It is important to know that indicators are not only used to measure activities, but also all log frame levels.

Example with IUCS team

Level	Objectively Verifiable Indicators
Goal	Mean ICU mortality per quarter (Wrote while this was intended indicators for the goal, due to limited resources, this indicator was not measured. The assumption of this data readily attainable was not correct
Result	Median incidence of central line- associated blood stream infections per 1000 catheter days (per quarter using NNIS definitions)
Objectives	1.1% of ICUs with a defined roster team member 2.1% of ICUs with record of consistence participation in project activities
Output	1. % ICUS with name executives and active participation from at least one senior executive (vice President level or equivalent) 2. %of ICUs in which preventable errors are recorded surfaced by staff 3. % of ICUS with a completed base line culture study
Activities	1. Number of units with culture survey response rate More than 60% 2. % IOUS participating CUPS conference call 3. % ICUS participating in CUPS coaching calls 4. Number of teams check-up forms submitted each month 5. % CPUS team meet monthly 6. % CPUS team with active Executive partners 7. % CPUS team implementing team work tools 8. % CPUS team reporting success 9. % CPUS team repotting barriers to success
Inputs	1. Number of returned and completed forms 2. Number of returned completed exposure tools 3. Number of downloaded CPUS manual and training too

Means of Verification (MoV) where information will be Advantages and Disadvantages of LFA available on the OVIs, and Assumptions

Means of Variation/Data collection

1. Goal
2. Outcomes
3. Activities

Means of Verification are the tools used and processes followed to collect the data necessary to measure progress. The data collected may be quantitative or qualitative. Qualitative data is more open-ended and often collected through interviews, focus groups and other qualitative data collection methods. Quantitative data is used to satisfy requirement of frequency in time and proportion such as, how many, how often, what proportion, and how much, and is largely collected through surveys.

Collection of information methods may be structured, semi-structured or non-structured.

Merits and Demerits of the main Data collection methods in the context of OVIs (objectively verifiable indicators) and Assumptions.

Surveys

A method to collect information and insights on a topic, often through paper or digital questionnaires.

Advantages	Disadvantages
1. Ability to be anonymous may elicit more truthful responses 2. Can collect information from a large group of people 3. Low costs 4. Quick responses 5. Respondents have time to consider their answers	1. May be difficult to gain access to groups 2. May be dangerous in certain VE contexts 3. Time-consuming to acquire enough data 4. Limited to smaller number of settings 5. For participant observation, participants may take time to trust the evaluator

Focus Group Discussions

A group of participants, usually from a similar background (age, gender, profession, etc.), are asked about their attitudes towards a project or topic. Focus groups discussions are carefully facilitated by a third party and usually consist of 6 to 10 participants.

Advantages	Disadvantages
1. Focused on specific topic such as a community issue. 2. Discussions generate a good number of opinions. 3. Provides opportunity to exchange ideas with others. Allows for immediate cross-check on information. 4. Collects detailed information about a specific topic from an expert. 5. Allows in-depth exploration of ideas that emerge from other methods.	1. Moderator bias: Moderator may hint at or state their personal biases, skewing the results. 2. Social desirability bias: participants may answer questions dishonestly in order to be viewed favourably by other participants (e.g. youth may answer differently if police are in the room with them). 3. Groupthink: participants may agree with the group just to minimize conflict. 4. Lack of anonymity: participants may not share some sensitive information (e.g. a youth may not be comfortable sharing about their interaction with a VE group.

Key Informant Interviews

A one-to-one interview; can be structured, semi-structured, or unstructured. Interviewees may include government officials, NGO staff, private sector representatives, religious leaders, academics.

Advantages	Disadvantages
1. Collects detailed information about a specific topic from an expert 2. Allows in-depth exploration of opinion from other methods	1. Findings may be biased: requires interviews with large number of respondents with varied views 2. Can be difficult to identify "experts" 3. Can be difficult to schedule 4. Time-consuming

Observation: Two types of observation:

1. Non-participant observation is a method in which the observer has no direct contact with the subject.
2. Participant observation takes place when an observer engages directly with the subject of their observations (e.g. joining a population or organization) and participates in the activities that are being studied.

Advantages	Disadvantages
1. Can be used to identify why an intervention is not working well. 2. Can be used when difficult to apply formalized data collection methods, such as when a population is resistant to being interviewed. 3. May be able to observe events that drive VE, such as sermons that advocate violence.	1. May be difficult to gain access to groups 2. May be dangerous in certain VE contexts 3. Time-consuming to acquire enough data 4. Limited to smaller number of settings 5. For participant observation, participants may take time to trust the evaluator

9

Introduction to Impact Assessment

Introduction

Impact studies isolate the effect of an intervention by assuring that there is clean comparison between a treatment group that received the intervention and a comparison group that is just the same except that it did not get the intervention.

Impact studies test whether one thing causes another. Just knowing scores have gone up following an intervention doesn't mean improvement is attributable to that intervention.

The data are generated from an impact study used to identify best practices, to replicate successful programs or policies, and to avoid mistakes that might have negative implications for future projects.

Concept of Impact Assessment

Impact Assessment is a means of measuring the effectiveness of organizational activities and judging the significance of changes brought about by those activities. It is neither Art nor Science, but both. Impact assessment is intimately linked to Mission, and, in that sense, ripples through the organization.

Impact analysis is the process of predicting the consequences a disruption in business function might cause and how teams can work jointly to fix these problems.

Concept of Impact Analysis

Impact analysis is the process of predicting the consequences a disruption in business function might cause and how teams can work together to fix these problems.

Meaning, concept and purpose in different context

Meaning

What is meant by impact assessment?

Impact assessment (IA) is a structured a process for considering the implications, for people and their environment of proposed actions while there is still an opportunity to modify (or even, if appropriate, abandon) the proposals.

Purpose in different context

An impact assessment is a planning and decision-making tool used to assess the potential positive and negative effects of proposed projects. Impact assessments consider a wide range of factors and propose measures to mitigate projects' adverse effect.

Impact assessment Framework

Important point of impact assessments includes global assessments (global level), policy impact assessment (policy level), strategic environmental assessment (program and plan level), and impact assessment (project level).

An impact evaluation provides prediction about the observed changes or 'impacts' produced by an intervention. These observed changes can be positive and negative, intended and unintended, direct and indirect. An impact evaluation must establish the cause of the observed changes.

An impact measurement framework builds on theory of change by identifying the data we will collect to understand, evaluate and demonstrate impact. It consists of measures that our organization will use to evaluate the tangible change that arises from our activities.

Main Steps of the Impact Assessment Process

Step 1: Select the Project(s) to be assessed

Step 2: Conduct an Evaluability Assessment

Step 3: Prepare a Research Plan

Step 4: Contract and Staff the Impact Assessment

Step 5: Carry out the Field Research and Analyse Results

Step 6: Disseminate the Impact Assessment Findings.

Impact Assessment Framework

The Development impact aims to define more precisely and formalise in a simple, efficient and systematic manner, in the project evaluation context, what "added value" or "doing good" means from an economic, environmental, and social and governance perspective as a contribution to the overall development impact. In practical terms, the project appraisal team will be required to provide a qualitative judgement based on statistics in seven crucial project performance aspects, closely linked with the potential development impact and, based on them, to reach a conclusion on the overall project rating in terms of development implication.

Performance Needs Considerations

Financial Performance: Financial success is a key and precondition for achieving development in all projects. The direct impact is related to project costs actually incurred (sunk costs) and the indirect impact to the negative image of a project failure both to the promoter and the country. Indeed, bankrupt projects have cost several developing countries more than a waste of valuable and scarce resources, it has cost them their reputation

Economic Performance: Unless a project has a sustainable economic performance, it will imply a waste of resources and consequently a negative development impact.

Economic viability depends upon the sustainability of project effects. Projects are sustainable if their net benefits or positive effects endure as expected throughout the life of the project. Sustainable development is concerned also with distributional issues. When looking at the distribution of project effects and judging project social acceptability, it is important to determine who benefits and who pays the costs. Job creation

Social Performance: Distribution analysis is also related to the social performance of the project. It is increasingly recognized that people are the centre of development, and that development should be for all people. The concept of social dimensions captures the key elements of human perspective: including poverty reduction; enhancing the role of women in development; human resources development; and avoidance or mitigation of adverse impacts of development interventions on vulnerable groups which do not have the capacity to absorb such effects. The social performance of a project is closely linked to community development. Local community participation in economic activity is another avenue for broadening the distribution of benefits of growth and for strengthening its sustainability.

Governance/Institutional Aspects: Governance and institutional aspects have become widely recognized as an important issue in developing countries. Weak governance discourages foreign investment in specific companies, reduces capital flows to developing economies in general, and can suppress information on a company's use of capital. Improvements to the investment climate, proper corporate governance, and support for firms that follow good practices contribute to development. In fact, recent research tends to demonstrate that there is a correlation between a company's approach to corporate governance and its risk profile, brand value, reputation, and ability to attract human and intellectual capital. The ability of a country to follow sustainable development paths is determined to a large extent by the capacity of its people and its institutions. The goal is to enhance the ability of a country or its principle economic actors to evaluate and address the crucial questions related to policy choices and modes of implementation among development options.

Environmental performance evaluation of a promoter and an International Funded (IF) project. Employment opportunities offering a safe, high-quality environment and medical care (including HIV/AIDS) make important contributions to reducing poverty and improving the quality of people's lives.

What is Research Impact?

Research impact is the contribution that our research makes to the advancement of knowledge, policy, practice, or society. It can be measured in different ways, such as by citations, publications, awards, media coverage, patents, or policy changes. However, these indicators do not capture the full range of impacts that our research can have, such as improving health, education, environment, or social justice. Therefore, we need to use more comprehensive and contextualized methods to assess the impact of our research.

Why is Impact Assessment Important?

Impact assessment is important for several reasons. First, it can help you clarify your research purpose and objectives, and align them with the needs and expectations of your target audience. Second, it can help you design your research methods and data collection strategies, and ensure that they are relevant, ethical, and feasible. Third, it can help you analyze and interpret your data, and identify the outcomes and impacts that your research has achieved or influenced. Fourth, it can help you communicate your research results and recommendations, and demonstrate your credibility and accountability. Fifth, it can help you improve your research quality and impact, and learn from your successes and challenges.

How to Choose an Impact Assessment Tool or Framework?

When choosing an impact assessment tool or framework for our research project, there are several factors to consider, such as the scope, stage, and duration of the project; the type and level of evidence needed; and the resources and expertise available. Common tools and frameworks include logic models, theories of change, impact evaluations, case studies, and impact stories. Logic models can help to plan, monitor, and evaluate research project and its impact, theories of change can help to articulate our impact pathway; impact evaluations can help measure causal effects; case studies can illustrate the complexity of research impact; and impact stories can capture the human aspects of research impact.

Use of Impact Assessment Tool or Framework

Using an impact assessment tool or framework effectively requires following general steps. First, we need to define our research impact question and objectives. Then, select the most suitable tool or framework to answer our research impact question and objectives. Afterwards, design an impact assessment plan that considers ethical and practical issues. After that, implement the plan by determining when and how we will conduct data collection and analysis, and who will be consulted. Finally, communicate our impact assessment results by presenting and disseminating findings and recommendations to the relevant parties.

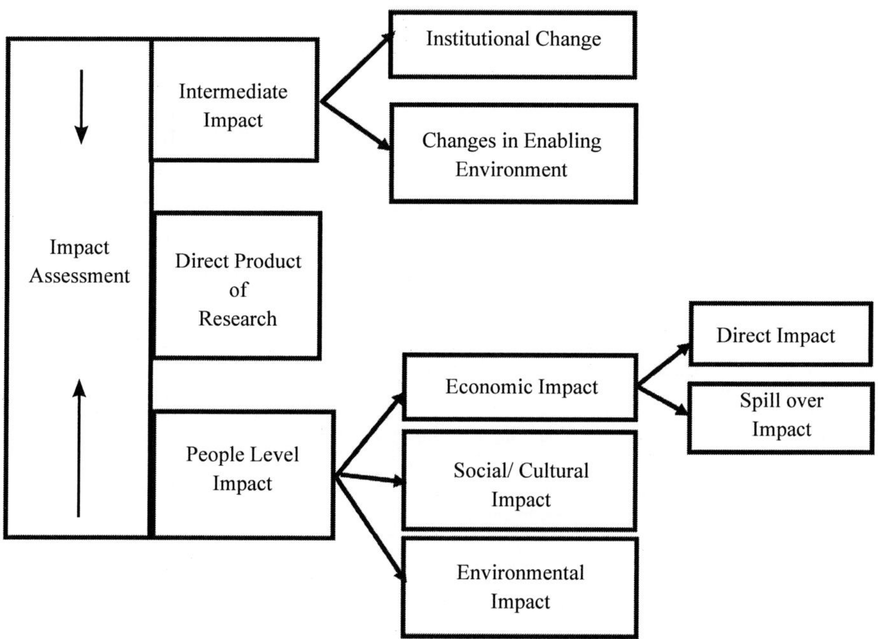

Planning and Managing Include

i) Describing what needs to be evaluated and developing the evaluation in brief.
ii) Identifying and mobilizing resources.
iii) Deciding who will conduct the evaluation and engaging the evaluator(s).
iv) Deciding and managing the process for developing the evaluation methodology.
v) Managing development of the evaluation work plan.
vi) Managing implementation of the work plan including development of reports.

Benefits and Challenges of Impact Assessment

Impact assessment can bring many benefits to our research project and its impact, such as enhancing relevance and quality, increasing visibility and recognition, strengthening collaboration and engagement, supporting funding and sustainability, and informing learning and improvement. However, there are also some challenges associated with impact assessment, such as defining and measuring impact in a clear and consistent way, establishing causality between research and impact, balancing rigour and feasibility in data collection and analysis, addressing ethical and political issues, and managing expectations among stakeholders. It is important to be aware of the benefits and challenges of impact assessment, as well as adopt a flexible and reflective approach to conducting it.

Meaning of Inputs, Outputs, Outcomes Impacts and their Relation with Monitoring and Evaluation

The end point is output. It is to be delivered. These are either tangible or intangible. These come from product related activities. Their relationship among and between inputs, activities, output, out come and impact.

Terms	Meaning
1. Inputs	The financial, human, and material resources used for the development intervention.
2. Activity	Activities designed to contribute pertinent environmental information to project or programme decision making. a process which attempts to identify, predict and assess the likely consequences of proposed development activities
3. Output	Outputs: the tangible and intangible products that result from project activities. Results chain: a graphical representation of the hypothesized relationship between project inputs, activities, outputs, outcomes and impact
4. Outcomes	Outcomes are the changes or benefits that result from the activities being taken or the outputs existing (e.g. changes in behaviour, knowledge, attitudes, competency, etc.) Impact is the long term or broader effect of the change (e.g. Changes in policy, systemic changes, etc.
5. Impact	Impact Assessment is a means of measuring the effectiveness of organizational activities and judging the significance of changes brought about by those activities

Evaluation and Impact Assessment

An impact evaluation provides information about the observed changes or 'impacts' produced by an intervention. These observed changes can be positive and negative, intended and unintended, direct and indirect. An impact evaluation must establish the cause of the observed changes.

An impact evaluation provides information about the impacts produced by an intervention.

10

Impact Assessment Indicators

Indicators for Impact Assessment-meaning and Concept

Indicators for impact assessment: Indicators are common tools to determine the performance and then the result of the technology/extension/communication interventions. They measure the accomplishment of the project's goals and targets. Criteria of indicators should be defined during the formulation stage, but they often need to be specified in greater detail during implementation. The indicators should measure intangible and non-tangible changes, particularly in projects that value factors such as personal and social development.

Impact assessment focuses on the effects of the intervention, whereas evaluation is likely to cover a wider range of issues such as the appropriateness of the intervention design, the cost and efficiency of the intervention, its unintended effects and how to use the experience from this intervention to improve.

There are eight leading principles that govern the entire Evaluation Impact Assessment (EIA) process, namely participation, transparency, security, accountability, credibility, cost efficiency, flexibility and practicality.

Indicators for Impact assessment is to be decided taking objectives, stakeholders, environmental variables and inputs applied. Therefore, indicators are different for different projects. The constant factors would be beneficiaries and benefits obtained from the project.

Types of Impact Indicators for Technology and Extension Advisory Services

There are various types of indicators used in monitoring and evaluation, including input, output, outcome, impact, efficiency, effectiveness and sustainability indicator.

Types of Common Indicators	1. Input
	2. Output
	3. Outcome
	4. Impact
	5. Efficiency
	6. Effectiveness
	7. Sustainability

Technology and Extension services go together. Once we decide the technology has to be adopted and then seek different methods of extension for diffusion of the technology. Therefore, good characteristics of usable technologies are to be decided as the indicators like Relative advantages, Compatibility, Complexity, Triability, Divisibility and Marketability etc. The recognized key stakeholders are to be find out who have direct and indirect relationship of the project selected for Impact Assessment. Normally these are as mentioned.

Key Stakeholders	Role of Stakeholders
1. Research institutes 2. Extension organizations 3. Academic institutions/Universities 4. Input dealers 5. Rural communities 6. Grassroots Administrative Units 7. Farmer organizations	1. Implementation 2. Funding 3. Problem analysis 4. Training and capacity Development 5. Planning 6. Resource management

The research type, impact indicators and key methods used in Impact assessments are as follows.

Research type Direct research output	Key Impact Indicators	Key Methods and Tools
Technology Development	Achieved scientific products compared to expected output. Improved agricultural technologies.	1. Documentary research 2. In- depth interviews with key researcher and research institutions & with locally extension staff
Capacity Building	Changes in capacity research organizations Publication, training and Academic degree attainment	1. Observation 2. Documentary research 3. In- depth interviews with key researcher and research institutions & with locally extension staff

Institutional Impact	Changes in shocks, natural disasters, health, Economic change, Trend in migration Seasonal price and employment, health	1. Observation 2. Documentary research 3. In- depth interviews with key researcher and research institutions & with locally extension staff 4. FGD with stakeholders

Social and Behavioural Indicators, Socio-cultural Indicators, Technology Level in Environmental Impact Assessment Indicators

Social Impact indicators used to measure impacts

Sl. No.	Social Impact Indicators
1	Risk assessment
2	Economic impact
3	Cultural Impact
4	Environment Impact
5	Project program and policy assssment
6	Hazards

Behavioural Indicators of Impact Assessment

Behavioural indicators are traits or behaviours that employers and managers can use to assess the competencies of their employees. Competencies are the key abilities necessary to perform certain job roles, and leaders can often use behavioural indicators to determine the presence or absence of competency.

We have listed the areas of behavioural Indicators like,

1. Team work
2. Communication skill
3. Leadership
4. Conflict Management
5. professional expertise
6. Ethical sense and values
7. Organizational skills
8. Decision making skill
9. Passion and commitment

Behavioural indicators are often important for employers because they give them the tools to create an effective workforce. They may give leaders the ability to predict competency in their employees, which can help them quickly to find the most qualified candidates. It may also help them to avoid hiring mistakes and costs associated with recruiting, training and on boarding replacement staff members.

These are some important competencies that employers may seek in their employees with their corresponding positive behavioural indicators.

Indicators of Team Work Behavioural Indicators

This is an important competency for many employees, especially those who commonly work in collaboration with others. Teamwork can help employees craft solutions to workplace challenges and create a healthy office environment. These are some positive behavioral indicators that can indicate teamwork competency in an individual:

Sl. No	Indicators
1	Values other people's opinions
2	Enjoys the company of other staff members
3	Willing to take time to answer questions or provide help to others
4	Seeks help from other staff members
5	Keeps communication channels open
6	Responds well to constructive criticism
7	Considerate of other employees' feelings
8	Willing to change their mind when faced with new ideas
9	Consistently contributes without leaving work to others
10	Helps other staff members complete their work

Indicators of Communication Skill

Communication is often a vital trait in the workplace, and many employers may look for this competency in their employees. It can allow staff members to avoid conflict, build working relationships and contribute to their team. Here are some positive indicators of strong communications skill.

Sl. No.	Communication Skill Indicators
1	Ability to clearly express ideas to others
2	Active listener to other employees
3	Interested to receive feedback well from others
4	Capable of processing feedback in proper perspective.
5	Possession of strong communication skills

6	Ability to resolve conflict through communication
7	Ability to communicate through writing and verbal communication
8	Able to confront others with challenging information without causing conflict
9	Holds the attention of others when communicating

Behavioural Indicators of Leadership

Leadership is an important competency for managers, directors and other personnel that oversee staff. It can help these individuals inspire trust in their teammates and direct junior employees towards company goals. These are some behavioral indicators that employers or managers might seek when searching for potential leadership candidates:

Sl. No.	Leadership Indicators
1	Motivates those around them to excel
2	Inspires trust in co-workers
3	Enjoys making decisions with others
4	Comfortable in taking responsibility and accepting consequences
5	Takes on mentoring and advisory roles with co-workers
6	Recognizes the achievement of others
7	Possesses strong sense of company goals
8	Possesses the ability to think strategically
9	Displays competence
10	Concerned for helping other staff members succeed

Behavioural Indicators of Conflict Management Skills

Many teams occasionally experience conflict and it's important to find employees who can navigate these challenging moments. These employees can help maintain a healthy team dynamic and a pleasant workplace for their co-workers. These are some behavioral indicators that employers can use to gauge their employees' conflict management competency.

Sl. No.	Indicators
1	Avoids conflict when unnecessary
2	Able to tell others when their behavior is inappropriate
3	States their opinion even when difficult
4	Listens to and considers the views of others
5	Handles conflict in an ethical way

6	Successfully diffuses confrontations
7	Attempts to resolve conflict in person before contacting a supervisor
8	Has the ability to find solutions without bias
9	Ability to remain calm in heated situations
10	Knows when to seek the help of supervisors or outside parties

Behavioural Indicators of Professional Expertise

This competency involves excellence in the technical aspects of a job. This can include many different skills depending on industry, but the indicators of expertise are often similar. These are some general indicators that an employee possesses expertise in their work:

Sl. No.	Indicators
1	Volunteers to help others to perform their work
2	Inspires trust in supervisors
3	Professionals be first person for their coworkers ask for help
4	Answers job-related questions directly, quickly and clearly
5	Accomplishes tasks on time
6	Rarely asks for help
7	Requires little supervision from managers
8	Willing to take extra responsibilities
9	Ability to clearly explain their job and responsibilities
10	Displays confidence at work

Behavioural Indicators of Ethical Sense and Values

This competency allows employers to trust the decisions of their employees. It implies that they adhere to a strict code of behavior in their personal and professional life. These are some indicators of ethical competency:

Sl. No.	Indicators
1	Displays honesty in interactions even when difficult
2	Holds themselves and others to a high standard
3	Admits failures and uses them to learn and improve
4	Treats others with respect
5	Embraces conflict when necessary
6	Feels a responsibility to report unethical behavior
7	Leads others by example

8	Works as hard as or harder than others
9	Makes sacrifices in their personal life to help others
10	Is trustworthy and dependable

Behavioural Indicators of Organizational Skills

Organizational skills are a common requirement for professionals in many fields. They can help employee's complete large and diverse workloads effectively as well as maintain records and files. Here are some behavioural indicators of strong organizational skills

Sl. No	Indicators
1	Ability to complete different tasks simultaneously
2	Keeps a clean and ordered workspace
3	Creates classifications for different items
4	Locates work materials and files easily
5	Completes assignments on time
6	Keeps a regular routine
7	Usually arrives to work early or on time
8	Creates lists to organize their thoughts and tasks
9	Uses calendars and planners
10	Makes plans ahead of time

Behavioural Indicators of Decision Making

Decision-making is often an important competency for leaders and managers. These employees may have to make choices that impact their company, and employers often want to know that they can reliably make wise decisions. Here are some of the traits that individuals with strong decision-making skills may have:

Sl. No.	Indicators
1	Displays trust in themselves
2	Doesn't hesitate when faced with difficult choices
3	Takes accountability for their mistakes
4	Makes cost-benefit analysis
5	Is unafraid of making mistakes
6	Acts quickly after deciding
7	Quickly perceives meaningful patterns

8	Knows when to seek help from experts
9	Ability to evaluate abstract concepts
10	Adapts well to their environment

Behavioural Indicators of Passion and Commitment

Passion and a commitment to excellence are often common traits of successful employees. They can help companies find staff who value their careers and invest time and effort into their job performance. Employers who value these traits might look for these indicators:

Sl. No.	Indicators
1	Displays persistency even when facing challenges
2	Seeks helps when needed
3	Sacrifices personal needs to help the organization
4	Works late or difficult hours
5	Works to improve job performance outside of work
6	Praises the work of others
7	Displays a desire to stay in their job
8	Builds strong professional relationships at work
9	Shows creativity in helping improve the organization
10	Shows self-discipline

Socio-Cultural Indicators of Impact Assessment

Socio-cultural impacts refer to the ways in which tourism changes community and individual values, behaviour, community structure, lifestyle and overall quality of life; in relation to both the destination and the visitor.

Sl. No.	Indicators
1	Food security
2	Poverty reduction
3	Status of women improved
4	Distribution of benefits across gender and geographical locations
5	Changes in resource allocation
6	Changes in cash requirement
7	Changes in labour distribution
8	Nutritional implications

Environmental indicators of Impact Assessment

Sl. No.	Indicators
1	Climate change - CO2 and greenhouse gas emission intensities
2	Layer - ozone depleting substances
3	NOx emission intensities
4	Municipal waste generation intensities
5	Water treatment connection rates
6	Use of water resources

Institutional Impact Indicators, Social Science Indicators, Agriculture Extension Education Assessment Indicators

Institutional Impact Indicators

Sl. No.	Indicators
1	Changes in organizational structure
2	Number of scientists on change
3	Change in the research team
4	Multidisciplinary approaches and improvements
5	Changes in funding allocated to the program
6	Changes in participation both public and private
7	New techniques or methods

Social Science Indicators

Sl. No	Indicators
1	Changes in awareness
2	Change in knowledge, skills level of beneficiaries
3	Increases in the number of people reached
4	Policy changes
5	Changes in behavior, e.g., adoption
6	Changes in community capacity
7	Changes in organizational capacity (skills, structures, resources
8	Increases in service usage.

Extension Service Indicators

Sl. No.	Indicators
1	Increases in the number of people reached
2	Policy changes;
3	Changes in behaviour
4	Adoption
5	Changes in community capacity
6	Changes in organizational capacity (skills, structures, resources)
7	Increases in service usage

11

Approaches for Impact Assessment

Impact Assessment Approach-Quantitative, Qualitative Participatory and Mixed Methods with Their Advantages and Disadvantages

i) Quantitative research approach
ii) Qualitative research approach
iii) Participatory research approach
iv) Mixed method research approach

Advantages of Quantitative Research

Quantitative Research is Concerned with Facts & Verifiable Information

Quantitative research is primarily designed to capture numerical data, often for the purpose of studying a fact or phenomenon in their population. This kind of research activity is very helpful for producing data points when looking at a particular group, like a customer demographic. All of this helps us to better identify the key roots of certain customer behaviours.

Quantitative Research can be Done Anonymously

Unlike qualitative research questions, which often ask participants to divulge personal and sometimes sensitive information, quantitative research does not require participants to be named or identified. As long as those conducting the testing are able to independently verify that the participants fit the necessary profile for the test, then more identifying information is unnecessary.

Quantitative Research Processes don't need to be directly Observed

Whereas qualitative research demands close attention be paid to the process of data collection, quantitative research data can be collected passively. Surveys, polls, and other forms of asynchronous data collection generate data points over a defined period of time, freeing up researchers to focus on more important activities.

Quantitative Research is Faster than other Methods

Quantitative research can capture vast amounts of data far quicker than other research activities. The ability to work in real-time allows analysts to immediately begin incorporating new insights and changes into their work -

dramatically reducing the turn-around time of their projects. Less delays and a larger sample size ensures us will have a far easier go of managing our data collection process.

Quantitative Research is verifiable & can be used to Duplicate Results

The careful and exact way in which quantitative tests must be designed enables other researchers to duplicate the methodology. In order to verify the integrity of any experimental conclusion, others must to replicate the study on their own. Independently verifying data is how the scientific community creates precedent and establishes trust in their findings.

Disadvantages of Quantitative Approach

Limited to Numbers and Figures

Quantitative research is an incredibly precise tool in the way that it only gathers cold hard figures. This double-edged sword leaves the quantitative method unable to deal with questions that require specific feedback, and often lacks a human element. For questions like, "What sorts of emotions does our advertisement evoke in our test audiences?" or "Why do customers prefer our product over the competing brand?", using the quantitative research method will not derive a meaningful answer.

Testing Models are more Difficult to Create

Creating a quantitative research model requires careful attention to be paid to your design. From the hypothesis to the testing methods and the analysis that comes after.

Tests can be Intentionally Manipulative

Bad actors looking to push an agenda can sometimes create qualitative tests that are faulty, and designed to support a particular end result. Apolitical facts and figures can be turned political when given a limited context. We can imagine an example in which a politician devises a poll with answers that are designed to give him a favourable outcome - no matter what respondents pick.

Results are Open to Subjective Interpretation

Whether due to researchers' bias or simple accident, research data can be manipulated in order to give a subjective result. When numbers are not given their full context, or were gathered in an incorrect or misleading way, the results that follow cannot be correctly interpreted. Bias, opinion, and simple mistakes all work to inhibit the experimental process and must be taken into account when designing your tests.

More Expensive than other forms of Testing

Quantitative research often seeks to gather large quantities of data points. While this is beneficial for the purposes of testing, the research does not come free. The grander the scope of your test and the more thorough you are in it's methodology, the more likely it is that you will be spending a sizable portion of your marketing expenses on research alone.

Qualitative Approach

Advantages

It becomes possible to understand attitudes

Because of intimacy between interviewer and interviewee it becomes to know and study the attitude of the interviewee. This is the great advantage in case of qualitative research approach.

It is a content generator

It permits for genuine ideas to be collected from specific socioeconomic demographics. These ideas are then turned into data that can be used to create valuable content which reflects the brand messaging being offered.

It can provide insights that are specific to an industry

The qualitative research process uses a smaller sample size than other research methods. This is due to the fact that more information is collected from each participant. Smaller sample sizes equate to lower research costs. It permits to save money, time and quick to draw conclusions.

It can provide insights that are specific to an industry

Relationships and engagement are the two most important factors for customer retention. Modern brands can use qualitative research to find new insights that can further these two needed items so their communication to their core demographics is as accurate and authentic as possible.

It allows creativity to be a driving force

It wants observations instead of creativity. The qualitative research process goes in a different direction than traditional research. This format eliminates the bias that tends to come through collected data as respondents attempt to answer questions in a way that please the researcher. Their creativity becomes a commodity.

It is a process that is always open-ended

The qualitative research process allows researchers to get underneath these habits to mine the actual data that someone can provide. It accesses the emotional data that drives decision-making responses. Because it is an open-ended process, there is no "right" or "wrong" answer, which makes data collection much easier.

It incorporates the human experience

Facts are important. Statistics can identify trends. Yet, the human experience cannot be ignored. The human experience causes two different people to see the same event in two different ways. By using qualitative research, it becomes possible to incorporate the complexity of this type of data into the conclusions that come from the collected research.

It has flexibility

There isn't a rigid structure to the qualitative research process. It seeks authentic data and emotional responses instead. Because of this flexibility, trained researchers are permitted to follow-up on any answer they wish to generate more depth and complexity to the data being collected. Unlike research formats that allow for zero deviation, the qualitative research can follow any thought tangent and mine data from the answers provided.

It offers predictive qualities

People who have similar perspectives will have similar thought patterns. They may even purchase similar products. The data which is gathered through qualitative research is perspective-based, which is why it has a predictive quality to it.

It allows for human instinct to play a role

Qualitative research recognizes human instinct which is absent in other approaches.

It can be based on available data, incoming data, or other data formats

The qualitative research method does not require a specific pattern or format for data collection. Information reporting is based on the quality and quantity of information that is collected. If researchers feel like they are not generating useful results from their efforts, they can change their processes immediately. There are more opportunities to gather new data when using this approach.

It allows for detail-orientated data to be collected

Numerous restrictions are part of the data-collection process in most research methods. This is done to help create measurable outcomes in a short time period. Instead of focusing on a specific metric, qualitative research focuses on data subtlety. It wants as many details as possible, whether those details fit into a specific framework or not. It is within those details that genuine insights tend to be found.

Disadvantages of Qualitative Research

It is not a statistically representative form of data collection

The qualitative research process does not provide statistical representation. It will only provide research data from perspectives only. Responses with this form of research cannot usually be measured. Only comparisons are possible. and that tends to create data duplication over time.

It relies upon the experience of the researcher

The data collected through qualitative research is dependent upon the experience of the researchers involved in the process. Industry-specific data must be collected by a researcher that is familiar with the industry. Researchers must also have good interviewing skills, have the courage to ask follow-up questions, and be able to form professional bonds with participants to ensure the accuracy of the data.

It can lose data

Data must be recognized by the researchers in qualitative research for it to be collected. That means there is a level of trust present in the data collection process that other forms of research do not require. Researchers that are unable to see necessary data when they observe it will lose it, which lessens the accuracy of the results from the qualitative research efforts.

It may require multiple sessions

The qualitative research may be effective in collecting authentic data, but the small sample size of the research can be problematic. To make an important decision, numerous perspectives are often required to avoid making a costly mistake. It may need two to three sittings with respondents to complete data collection.

It can be difficult to replicate results

Because qualitative research is based on individual perspectives, it is almost impossible to duplicate the results that are found. Even the same person may have a different perspective tomorrow than they had today. That means the data collected through qualitative research can be difficult to verify, which can lead some to the conclusions that researchers generate through this process.

It can create misleading conclusions

Although like-minded people tend to think, feel, and act in similar ways, this is not always the case. There is no absolute way to know if the conclusions generated through qualitative research can apply to an entire demographic.

It can be influenced by researcher bias

The term "fake news" has been used quite often since the beginning of 2017. The term is reveals a certain bias that seems to be present in media reporting and the reporting is said to be unbiased. In such research, the bias of the researcher, whether conscious or subconscious, can affect the data.

It may not be accepted

There is a certain authenticity to qualitative research, there is also certain subjectivity to it. Because of this nature, the data collected may not be accepted.

It creates data that is difficult to present

Because individuals have different perspectives, the reaction to qualitative research findings can often be at two extremes. There will be those who support the findings and there will be people who do not support the findings. The data being collected will be viewed as valuable by both groups, but how each group chooses to act is on their own perspective. That means two very different outcomes can be achieved, making the data difficult to present to generalized audiences.

It creates data with questionable value

Even researchers may disagree about the value of data being collected because of their different perspectives. What is included during the qualitative research process or what is excluded relies upon the researcher involved. That is why this data collection process is highly subjective.

It can be time consuming

Because researchers follow numerous angles when collecting data, it takes more time to gather it. Sorting through all of that extra data takes time as well. Every data point is evaluated subjectively, so the worth of it is always in question.

It has no rigidity

The qualitative research method is based on individual perspectives. Since those perspectives can change, the data gathered is only reliable at the time it is gathered. Human memory tends to prefer remembering good things. We keep close access to fond memories and put bad memories into the back corner of the mind.

It lessens the value of data mining

Data mining can provide valuable insights to an entire demographic of customers. In qualitative research, data mining is useful for the one person who is providing information in the first place.

These key advantages and disadvantages of qualitative research show us that gathering unique, personalized data will always be important. It is the best method to understand how certain people, and even certain groups, think on a subject.

Advantages and Disadvantages of Participative Management

Advantages

Undoubtedly participative approach to management increases the stake or ownership of employees. But there is more to it. The following points elucidate the same.

Raise in Output: Participation in decision making encourage employees to be alert about their duties and responsibilities. Such encouragements result in higher output.

Employee Satisfaction: In participatory management system the employees feel elevated as their contribution is viewed as valuable. Such feelings bring satisfaction in the minds of the employees.

Interest and Motivation: More production and positiveness of the employee's are the output of motivational factors. Participatory management system creates environment of interest and motivation.

High Quality: Once the management keeps employees happy, they give more importance to quality for up keeping of name and fame of the business.

Lower Costs: Participatory management reduces cost of supervision because employees earn the skill of self-assessment consciousness about the organization in which they work. There is a lesser need of supervision and more emphasis is laid on widening of skills.

Demerits of Participative Management

Participatory management also not free of demerits.

Slow Decision: Because of more people and more participation the decision-making process becomes slow. It becomes difficult to reach final stage of decision making.

Protection Issue: It is natural environment that where more people are involved more problems arises to keep property of organization safe. In case of any loss none assume responsibility. This is the situation with participatory management system.

The advantages seem to outnumber the disadvantages. This however is no assurance that one should blindly adopt it for his/her organization.

Advantages and Disadvantages Mixed Method

Five purposes for mixed-method evaluations are identified in this conceptual framework: triangulation, complementarity, development, initiation and expansion. For each of the five purposes, a recommended design is also presented in terms of seven relevant design characteristics.

The pros and cons of mixed methods designs are important to take into consideration for this study.

Advantages

A Mixed Mode survey design often offers people the choice of joining that best suit their survey-taking preferences and resources. By reaching people through multiple survey modes, we can achieve better population coverage and response rates – and be more efficient in terms of time and money.

Strengths of this method: Quantitative methods are always attributed to natural science (positivist) whereas qualitative methods are often associated with social science.

Mixed methods bring out both quantitative and qualitative results of the study

Disadvantages

Needs more time to plan and complete

Difficult to interpret t findings

Research design becomes very complex and deficiencies are difficult to be handled.

It requires more resources for collection of both types of data (quantitative and qualitative).

Quantitative or qualitative results of the first phase of the study may fail to demonstrate significant differences for the second phase.

This method requires a consistent and clear presentation so that the audience can easily and accurately follow the procedures and the findings.

Researcher needs to be familiarized with multiple methods and displays the ability to mix each one of them effectively.

Methodological purists only believe in either qualitative or quantitative research. Alternatives methods not applicable.

Quantitative Impact Assessment Types based on Time of Assessment (Ex-ante and ex-post)

Ex-ante approach

Ex-ante Policy Impact Assessment (hereafter PIA) is an analytical process, conducted at the early stages of policy-making exercise. It comprises a set of logical steps to support the decision-making process by providing evidence-based policy alternatives, forecasting and listing their potential impacts.

PIA does not replace the decision-making process but rather informs it in a participatory manner, enriching the substantiation of the best policy option. This Quick Guide addresses the entire PIA process, which starts from problem definition, objectives' setting and options identification, informing the core step of PIA process, options impact assessment.

Therefore, there should not be confusion between comprehensive PIA, which is described in this Quick Guide and one of its stages impact assessments.

Mostly PIA is meant to answer two crucial questions:

1. Is the planed intervention justified?
2. Which are the best ways to solve the problems and to achieve the objectives?

PIA Steps Ex-ante Policy Impact Assessment includes five consecutive steps, namely:

1. Problem Definition
2. Objectives Setting
3. Identification of Policy Options
4. Impact Assessment of Policy Options
5. Comparison of Policy Options and Selection of the Best One

Before starting planning PIA, the following should be remembered

1. The sequence of these five steps is important, however PIA is an iterative process – findings generated at a certain PIA stage could suggest making a step back and revising the findings in the previous PIA stages.
2. PIA steps must be followed properly, with the level of analysis and tools used chosen on the capacity of the institution in some cases it will be sufficient answering precisely the questions inserted in each PIA step and use qualitative tools and when capacities are higher the quantitative tools will be used.
3. Notification of stakeholders at the beginning of PIA process and consultation throughout entire PIA exercise are critical (see details in the sections below).
4. Data-primary and secondary, qualitative and quantitative are key for better policy substantiation.
5. The whole PIA process should be always adjusted to the specific issue it is linked to. Thus, the depth and complexity of the PIA should be in line with the complexity of the issue.

The principle of proportionality is to be considered carefully for not wasting time and energy when that is not necessary. Overall, PIA should be considered necessary for the following types of public policies:

1. Public policies that introduce significant.
2. Public policies that relate to transfers to citizen.
3. CMU interventions in the business environment.
4. Long-term investment projects.

Based on Research Design (Experimental, Quasi Experimental, Non-Experimental) for Conducting Impact Assessment Assignment

What Is Experimental Research Design?

Experimental research design is a framework of protocols and procedures created to conduct experimental research with two sets of variables. Herein, the first set of variables acts as a constant, used to measure the differences of the second set. The best example of experimental research methods is quantitative research.

Experimental research helps a researcher gather the necessary data for making better research decisions and determining the facts of a research study.

When Can a Researcher Conduct Experimental Research?

A researcher can conduct experimental research in the following situations

1. When time is an important factor in establishing a relationship between the cause and effect.
2. When there is an invariable or never-changing behaviour between the cause and effect.
3. Finally, when the researcher wishes to know the importance of the cause and effect.

Importance

To publish significant results, choosing a quality research design forms the foundation to build the research study. Moreover, effective research design helps establish quality decision-making procedures, structures the research to lead to easier data analysis, and addresses the main research question. Therefore, it is essential to cater undivided attention and time to create an experimental research design before beginning the practical experiment.

By creating a design of research, a researcher is also giving oneself time to organize the research, set up relevant boundaries for the study, and increase the reliability of the results. Through all these efforts, one could also avoid inconclusive results. If any part of the research design is flawed, it will reflect on the quality of the results derived.

Types of Experimental Research Designs

Based on the methods used to collect data in experimental studies, the experimental research designs are of three primary types:

Pre-experimental Research Design

A research study could conduct pre-experimental research design when a group or many groups are under observation after implementing factors of cause and effect of the research. The pre-experimental design will help researchers understand whether further investigation is necessary for the groups under observation.

Pre-experimental research is of three types

One-shot Case Study Research Design

One-group Pretest-posttest Research Design

Static-group Comparison

True Experimental Research Design

An experimental research design relies on statistical analysis to prove or disprove a researcher's hypothesis. It is accurate forms of research because it provides specific scientific evidence. Furthermore, out of all the types of experimental designs, only a true experimental design can establish a cause-effect relationship within a group. In an experiment, a researcher must satisfy these three factors.

There is a control group that is not subjected to changes and an experimental group that will experience the changed variables.

A variable to be manipulated by the researcher.

Random distribution of the variables.

This type of experimental research is commonly observed in the physical sciences.

Quasi-experimental Research Design

The word "Quasi" means similarity. A quasi-experimental design is similar to a true experimental design. However, the difference between the two is the assignment of the control group. In this research design, independent variable is manipulated, but the members of a group are not randomly assigned. This type of research design is used in field settings where random assignment is either irrelevant or not required.

The classification of the participants, conditions, or groups determines the type of research design to be adopted.

Advantages of Experimental Research

Experimental research allows to test your idea in a controlled environment before taking the research to clinical trials. Moreover, it provides the best method to test theory because of the following advantages:

Researchers have firm control over variables to obtain results.

The subject does not impact the effectiveness of experimental research. Anyone can implement it for research purposes.

The results are specific.

Post results analysis, research findings from the same dataset can be repurposed for similar research ideas.

Researchers can identify the relation of cause and effect and further analyse this relationship to determine in-depth ideas.

Experimental research makes an ideal starting point. The collected data could be used as a foundation to build new research ideas for further studies.

Mistakes to Avoid

Invalid Theoretical Framework

Inadequate Literature Study

Insufficient or Incorrect Statistical Analysis

Undefined Research Problem

Research Limitations.

Ethical Implications.

Non-Experimental Research Design for Impact Assessment

Non-experimental designs may include pre- and post-intervention studies with no control or comparison group, case study approaches, and post-intervention-only approaches, among others. The key feature of a non-experimental design is the lack of a control group. Non-experimental research is research that is not subjected to the manipulation of an independent or predictor variable. Predictor variables are used to anticipate the outcome of other variables; they are the portion of the experiment that is manipulated to know the variation on the dependent variable. Instead, the research is conducted by observing and measuring non-manipulated variables in the context in which they occur and investigating them to obtain information. Variables in non-experimental research cannot be controlled, manipulated, or altered-therein the researcher has to try to objectively interpret the observational data to get a conclusion. Also, in case of only one group, there will not be experimental or control groups, so it would have non-experimental design. The aim of non-experimental research is to define the characteristics of a particular subject(s), measure data trends, compare and contrast situations, and validate existing conditions.

Non-experimental research designs can be :

i) Cross-sectional,
ii) Longitudinal,
iii) Correlation,
iv) Or observational.

Its main purpose is to be descriptive or describe situations or relationships. The researcher chooses to be descriptive, observational, find a correlation, or even use surveys as the method of data collection.

Econometric Impact Assessment (Partial budgeting technique, Net Present Value

The purpose of an economic impact assessment is to estimate the changes in employment, income, and levels of business activity (typically measured by gross receipts or value added) that may result from a proposed project or program.

4 Stages of Econometrics

1st Stage	Develop a theory or hypothesis. Econometricians first establish a hypothesis or theory to guide data analysis
2nd Stage	Specify a statistical model. In this step, econometricians identify a statistical model to examine the relationship between variables
3rd Stage	Estimate the model's variables
4th stage	Perform a test

Partial Budgeting Technique

It is a planning and decision-making structure adopted to compare the costs and benefits of alternatives faced by a farm business. It determines changes in income and expenses that would emerge on implementing a specific alternative.

Thus, all aspects of farm profits that are unchanged by the decision can be safely ignored. In a nutshell, partial budgeting allows to get a better handle on how a decision will affect the profitability of the enterprise, and ultimately the profitability of the farm itself.

When and How to Use Partial Budgets

The partial budget framework can be adopted to examine a number of farm decisions, including:

i) Adopting a new technology
ii) Changing enterprises
iii) Choosing to specialize
iv) Hiring custom work

v) Leasing instead of buying machinery
vi) Modifying production practices
vii) Making capital improvements

The structure of the analysis depends upon the types of the decision being analysed. For example, in case we want to analyse the installation of a new milking parlour. It would be wise to perform a partial budget analysis on the milking enterprise by analysing costs and returns on either a per-cow or per-hundredweight basis. On the other hand, a farmer choosing to purchase feed rather than grow it might want to see the effects on farm, on total income and costs. The partial budgeting framework is flexible enough to allow for these modifications.

Keep in mind that partial budgeting analyzes the impacts of some change on profit. Prior analysis should be performed to assure that the enterprise or farm, whichever is being analysed, is profitable. If it is not profitable, we may have much more important decisions to face.

There are seven steps to the successful use of partial budget analysis as a decision-making tool. (Partial Budgets, each step serves a specific, unique purpose and is vital to an accurate, meaningful analysis).

Step 1: State the proposed change

It is important to have a clear understanding of exactly what alternative is being analyzed. If possible, we should analyse several alternatives; however, the analysis of each is carried out on an individual basis. Clarity at this stage will help us to easily complete the other steps.

Step 2: List the added returns

Identify any possible means of generating new revenue streams or increasing existing streams. Suppose the alternative is purchasing a mixer. Will that lead to increased milk production? If so, then the added revenue resulting from growth in milk sales should be determined.

Step 3: List the reduced costs

In this step, begin by identifying general areas where the choice might lower expenses. Once all general areas are identified for the specific alternative, one can work to plug numbers into the partial budget. For example, the choice to hire custom crop harvesting. It is the most obvious savings associated with this decision is a decreased need for labour. We can identify how many hours of labour will be saved, and then multiply that figure by the hourly wage rate to obtain a value for the partial budget. This is one reduced cost. It is important to find all possible costs that the choice will reduce.

Step 4: List the added costs

Once again, start by identifying all of the general areas in which costs will be increased. The choice to have crops custom-harvested has one obvious new cost: the payment for the service. Suppose instead the choice was to purchase a new piece of machinery for $50,000 with a useful life of 10 years, a salvage value of $5,000, a financing interest rate of 7 percent, and repair and insurance rates of 2 percent and 3 percent of the average value, respectively. In the situation of capital purchases, a depreciated cost must be claimed annually, not the total purchase cost. The average value ($27,500) should be used to compute annual interest, repair, and insurance expenses. To summarize, the added annual costs for the purchase of this piece of machinery are:

Example:

1. Depreciation: $ 4,500
2. Interest: $ 1,925
3. Repairs: $ 550
4. Insurance: $ 825

Be careful to thoroughly analyze the alternative to get a handle on all sources of added costs.

Step 5: List the reduced returns

Will revenues be decreased or eliminated as a result of choosing a particular alternative? For instance, suppose you are deciding whether or not to adopt a no-till system of crop production. Will this decrease yield? If so, then you must estimate the amount of the reduction and multiply that by an expected price to approximate the reduced revenues resulting from the adoption of no-till.

Step 6: Summarize the net effects

Once we have identified the individual positive (steps 2 and 3) and negative (steps 4 and 5) aspects of the alternative, these should be aggregated to determine a total cost and total benefit of the alternative. The net benefit of the alternative is calculated by subtracting total costs from total benefits. The positive results indicates that alternative may have economic advantages. However, if the net benefit is negative, the business would be better off staying with the current situation or analysing a different alternative.

Step 7: Consider non-economic and other factors

Non-economic considerations in important to think of an alternative. Such considerations may include the social aspects of having less labour on the farm, increased/ decreased leisure time, the need for increased or specialized knowledge, and safety and/or ease of use of equipment. Note that these are generally focused on quality-of-life measures, which are frequently difficult to quantify.

As we work through the partial budget analysis, it is essential to know those numbers in the analysis that can be called "hard numbers." Hard numbers are the values we can assign with a higher level of certainty.

Suppose, we are concerned that the harvester is not be available at an optimal time. This may lead to decreased output. However, it is unclear exactly how much production might be lost. Thus, lost production can be considered a "soft number." In this situation, we should incorporate our best estimate in the partial budget analysis, then use different yield loss numbers to see how much the soft number estimate should change before the decision switches.

It can be beneficial to use "best-case" and "worst-case" numbers to establish a range for the partial budget analysis. Obviously, the fewer the soft numbers you use, the better. The softer numbers in the analysis, indicates less trustworthy of the results. This is why a good partial budget analysis must be founded upon good records, which provide many hard numbers.

Ensuring Realistic Estimates

A decision based on partial budget analysis is only as good as the information in the analysis will allow. We can ensure that we are using realistic and accurate figures for price savings and expenses in our analyses:

1. Review previous years' actual expenses
2. Use the Internet to research fees associated with services
3. Contact your Penn State Extension agent
4. Get prices from several suppliers
5. Talk with other producers who use the alternative you are considering

By speaking with farm managers or producers who already made a change similar to what you are considering, you can also learn about things they wish they had done differently, problems they encountered, or successes they achieved.

Net Present Value (NPV)

Net Present Value (NPV) is adopted to calculate the current value of a future stream of payments from a company, project, or investment. To calculate NPV, we need to estimate the timing and amount of future cash flows and pick a discount rate equal to the minimum acceptable rate of return.

Net present value (NPV) is Present Value of Cash inflow minus Present value of cash out flow over a time period (Value cash inflow - Value of cash out flow).

NPV is used in capital budgeting and investment planning to analyze the profitability of a projected investment or project.

NPV is the result of calculations that find the current value of a future stream of payments using the proper discount rate. In general, projects with a positive NPV are worth undertaking, while those with a negative NPV are not.

Key Points

1. NPV is used to calculate the current value of a future stream of payments from a company, project, or investment.
2. To calculate NPV, we need to estimate the timing and amount of future cash flows and pick a discount rate equal to the minimum acceptable rate of return.
3. The discount rate may reflect our cost of capital or the returns available on alternative investments of comparable risk.
4. If the NPV of a project or investment is positive, it means its rate of return will be above the discount rate.

Net Present Value (NPV) Formula

$$NPV = \frac{Rt}{(1 + i)t}$$

NPV = Net Present Value

Rt = Net cash flow at time t

i = Discount rate

t = Time of cash flow

The net present value or net present worth applies to a series of cash flows occurring at different times. The present value of a cash flow depends on the interval of time between now and the cash flow. It also depends on the annual effective discount rate. NPV accounts for the time value of money.

Indication of NPV

NPV accounts for the time value of money and can be used to compare the rates of return of different projects or to compare a projected rate of return with the hurdle rate required to approve an investment.

The time value of money is represented in the NPV formula by the discount rate, which might be a hurdle rate for a project based on a company's capital cost. No matter how the discount rate is determined, a negative NPV shows that the expected rate of return will fall short of it, meaning that the project will not create value.

In the context of evaluating corporate securities, the net present value calculation is often called discounted cash flow (DCF) analysis. It's the method used by Warren Buffett to compare the NPV of a company's future DCFs with its current price.

The discount rate is central to the formula. It accounts for the fact that, as long as interest rates are positive, a dollar today is worth more than a dollar in the future. Inflation erodes the value of money over time. Meanwhile, today's dollar can be invested in a safe asset like government bonds; investments riskier than Treasury's**1** must offer a higher rate of return. However, it's determined, the discount rate is simply the baseline rate of return that a project must exceed to be worthwhile.

Positive NPV vs. Negative NPV

A positive NPV indicates that the projected earnings generated by a project or investment-discounted for their present value-exceed the anticipated costs, also in today's dollars. It is assumed that an investment with a positive NPV will be profitable.

An investment with a negative NPV will result in a net loss. This concept is the basis for the net present value rule, which says that only investments with a positive NPV should be considered.

Benefit Cost Ratio Internal Rate of Return, Adoption Quotient

The benefit-cost ratio is used to determine the viability of cash flows from an asset or project. The higher the ratio, the more attractive the project's risk-return profile. Poor cash flow.

Benefit-Cost Ratio (BRC)

BCR is a ratio to indicate an analysis of cost and benefit along with relationship between the two. The outcome can be expressed quantitatively (money) and qualitatively (change) If a project has a BCR greater than 1.0, the project is expected to deliver a positive net present value to a firm and its investors.

Important Points

The BCR shows the relationship between the costs and benefits of a proposed project.

If a project has a BCR greater than 1.0, the project is expected to be profitable.

If a project's BCR is less than 1.0 it will not be profitable

How the Benefit-Cost Ratio (BCR) Works

BCR is used in capital budgeting to reveal positive and negative side, i.e. Profit or loss. It is in big business establishment. In case of large project uncertainties are more for which profitability calculation becomes difficult. BCR takes care of it.

The BCR also does not provide any sense of how much economic value will be created, and so the BCR is usually used to get a rough idea about the viability of a project and how much the internal rate of return (IRR) exceeds the discount rate, which is the company's weighted-average cost of capital (WACC) - the opportunity cost of that capital.

The BCR is unable to indicate creation of economic values. But can provide a rough idea about viability of the business or project. It also indicates how much Internal Rate of Return exceeds the discount rate. It is the weighted average cost capital (WACC).

"The BCR is calculated by dividing the proposed total cash benefit of a project by the proposed total cash cost of the project. Prior to dividing the numbers, the net present value of the respective cash flows over the proposed lifetime of the project taking into account the terminal values, including salvage/remediation costs – are calculated".

A project with BCR higher than 1.0 it is in positive side, if equal to 1.0 indicates profit equal to cost and if less 1.0 is negative side. The project has to be dropped.

Adoption Quotient

A.Q. is a ratio scale designed to quantify the adoption behaviour of an individual. To measure the adoption quotient potentiality, extent, time and consistency were considered. The numerical value as calculated from the formula was used for individual respondent.

Adoption is conceived as a 2-part process: (1) cognitive adoption, involving obtaining knowledge and critically evaluating the practices; and (2) behavioral adoption, involving actual uses of the practices. The variables involved in the 2-part process are identified and used to construct the adoption quotient (AQ), a ratio scale designed to quantify adoption behaviour of an individual. The validity of the AQ scale was tested on 10 high-AQ and 10 low-AQ farmers. The correlations obtained suggested that the scale is valid.

12

Environmental Impact Assessment

Concept of Environmental Impact Assessment Introduction, what it is, who does it, how it is done?

Environmental Impact assessment (EIA) is the estimation of the environmental consequences of a plan, policy, program, or actual projects prior to the decision to move forward with the proposed action.

Introduction to EIA

An EIA may be defined as: a formal process to predict the environmental consequences of human development activities and to plan appropriate techniques to eliminate or reduce adverse effects and to augment positive effects. to enhance positive effects.

EIA: A tool used to identify and assess the potential impacts of a proposed project (or activity), evaluate alternatives, and formulate appropriate mitigation, management and monitoring measures. predict the likely environmental impacts of the project.

What it is?	Environmental Impact Assessment (EIA) is done to determine what effects proposed projects and programs will have on aspects of the environment, including its human dimensions.
Who does it?	On 27 January 1994, the Union Ministry of Environment and Forests (MEF), Government of India, under the Environmental (Protection) Act 1986, promulgated an EIA notification making Environmental Clearance (EC) mandatory for expansion or modernization of any activity or for setting up new projects listed in Schedule foe EIA
Why EIA is done?	The goal of carrying out environmental assessments is to protect human health and the natural environment from the foreseeable, harmful effects of planned industrial facilities and infrastructure measures
How EIA is Conducted?	This process involves several steps, including screening to determine the level of detail needed, scoping to define the objectives and boundaries of the EIA, baseline to collect and analyze existing conditions in the project area, impact analysis to identify and predict the likely environmental impacts of the project

Benefits and Important Aspect of EIA, Risk Assessment, Environmental Assessment and Post Product Monitoring

i) **Benefits:** The benefit of having an EIA is to help prevent more severe environmental damage and pollution. In addition, the EIA project has benefits for the government; this project can be used as a form of government responsibility in preserving and protecting the environment.

ii) **Important Aspects of EIA:** EIA assesses a project's potential impacts on air quality, including emissions of pollutants. It also considers impacts on water resources, such as water availability and quality changes. Additionally, soil quality is assessed to determine potential impacts on agricultural productivity and soil erosion.

Risk Assessment: The Risk Assessment Report may cover the following in terms of the extent of damage with recourse to MCA (Multi criteria Analysis) analysis and delineation of risk mitigation measures with an approach to DMP (Damp Proof Membrane). Hazard identification: Identification of hazardous activities, hazardous materials, past accident records etc.

Types of Risk Assessment & How to Use Them

- Qualitative risk assessment
- Quantitative risk assessment
- Generic risk assessment
- Site-specific risk assessment
- Dynamic risk assessment

In basic terms, a risk assessment identifies risks and constraints in the workplace and determines methods of either mitigating or eliminating them altogether to ensure a safe working environment. As we know, hazards come in all shapes and size meaning risk assessments do too. To accurately evaluate the variety of hazards hidden in a workplace, we should use a different type of risk assessment. It's important to know that different types of risk assessments can often be used alongside one another. Workplace hazards aren't black and white, so sometimes we may need to mix and match, collating the most relevant parts of each type of assessment into one single risk assessment.

Stages of a Risk Assessment

While there are 5 different types of risk assessment, the stages involved in carrying out each assessment are very similar. A risk assessment should include these 6 steps:

Plan

Good planning is integral to ensuring to get our risk assessment off to a strong start. Before conducting risk assessment, consider these four elements noted by.

Identify Hazards

This requires participation of a walkthrough of our workspace to identify anything expected to cause harm. Involving other employees in this stage could be helpful, as they may have noticed things that don't immediately stick out at us.

While assessing the workplace, ensure we consider continued constraints and safety hazards.

Once we have identified each hazard, make sure that we understand how it could harm someone or the environment as this will help to identify the most impactful way of managing the risk further down the line.

Evaluate Identified Risks

Now that hazards have been identified, it's time to decide how to handle them.

Under the OSH Act, employers have a duty to provide workplaces that are free of known hazards that could harm their employees. Comparing what we are already doing with best practice is the most effective way to do this.

First, assess the controls we currently have in place and how they're implemented. Compare these controls against identified best practices to see if there's room for improvement.

Take Action

Once we have identified specific areas for improvement to reduce risks, it's time to act. HSE (Health, Safety, and Environment) suggests applying the following principles:

- Desirable is the less risky option better to switching of to a less hazardous chemical
- Prevent access to the hazard (such as guarding)
- Organize work to reduce the extent of hazard (for example, by placing barriers between pedestrians and traffic)
- Provide personal protective equipment (gloves, goggles, etc.)
- Provide emergency areas (like first aid or eye wash stations)

Record Findings and Actions Taken

At this stage, we should list out the results and share them across our organization to encourage all employees to put these new actions into practice.

Review Process

Workplaces seldom stay the same. New processes could be brought in that present new risks and hazards. Better to review the process on regular basis. Risk assessments come in many different forms, but here are the five most common:

Qualitative Risk Assessment

Qualitative risk assessments are the most frequently used risk assessments for companies in high-risk industries. Assessments of this kind measure the severity of a risk. In most cases, qualitative risk assessments are used to determine the severity of multiple risks at once. The severity of a risk can be determined using the following formula: impact describes how severely a risk could negatively affect a project and likelihood describes how often the risk is likely to occur.

Severity = Impact x Likelihood

When the information is presented, it is organized either into categories or a graph known as a Qualitative Risk Assessment Matrix. Both schemes are based on severity. Categories are labelled High, Medium, and Low or, depending on the company's preferences, Green, Amber, and Red (GAR). In a Qualitative Assessment Matrix, one axis is labelled 'Impact' and the other 'Likelihood,' and plot points (risks) are charted in according to severity. Presentations allow the company to quickly identify the highest-severity risks.

Quantitative Risk Assessment

A quantitative risk assessment involves determining the severity and possibility of a risk by giving it a number. This type of assessment aims to gauge how much the impact of the risk will cost an organization. Rather than referring to a risk as high, medium or low, a quantitative risk assessment assigns a digit to the risk.

The main difference between a qualitative and quantitative risk is that rather than being based on a person's judgment, as a qualitative risk is, a quantitative risk analysis relies on hard data. Another contrast between the two is how each process values risk. While in a qualitative risk assessment, an innocuous risk may be rated "low," the same risk will be given a low percentage in a quantitative assessment to indicate the possibility of it occurring or causing harm.

A scenario in which quantitative risk determinants are commonly used is to predict the likelihood of a fire or explosion when using harmful chemicals. Project managers can select one of the following methods for quantitative risk assessment:

1.	Failure Mode and Effects Analytics (FMEA) - anticipating shortcomings in business processes and coming up with ways to mitigate their impact on customers.
2.	Business Impact Analysis (BIA) - identifies and evaluates impact of natural disasters and sets aside investment in recovery, prevention and mitigation strategies for organizations effected.
3.	Expected Monetary Value (EMV) - this method involves senior management carrying out the following equation: Probability in % of Risk Occurring x Cost of Impact in Preferred Currency.

Generic Risk Assessment

Generic risk determination includes assessing a wide of specific workplace activities. Generic risk assessments are most useful regarding repetitive tasks which are consistent in application for organizations.

Site-Specific Risk Assessment

As the name suggests, a site-specific assessment places a particular site, environment and team of employees performing the task under the microscope to ensure everything is running as safely as it should be.

The main difference between generic and site-specific assessments is that while generic risk assessments place a spotlight on common hazards associated with general activities carried out regularly by employees, such as roofing works, site-specific risk assessments pay attention to hazards that can only be found in one specific location with controls put in place that are very distinct to the risk – for example, valley gutters on a roofing site.

Dynamic Risk Assessment

A dynamic risk assessment evaluates risks in rapidly changing, uncertain and often high-risk environments. In contrast to formal risk assessments, or legal requirement and should be carried out as a precursor or motivator to any task being done, dynamic risk assessments are normally carried out by a lone worker when they arrive in a new environment. There may also be circumstances where a person's regular working environment changes, making generic or even site-specific risk assessments obsolete and presenting a dynamic approach.

Environmental Assessment

Environmental assessment can be defined as identifying, estimating, and evaluating the environmental impacts of existing and proposed projects, by conducting environmental studies, to mitigate the relevant negative effects prior to making decisions and commitments.

Strategic Environmental Assessment (SEA)

SEA is a systematic process for evaluating the environmental consequences of proposed policies, plans, or programs in order to include them in policy making on par with economic and social considerations. SEA is often considered to be complementary to project-based EIA, but is also an evolved paradigm of environmental assessment moving EIA principles upstream in the policy-making processes. The international regulatory framework for SEA is mainly set by two legal documents, the European SEA Directive and the United Nations Economic Commission for Europe but several other SEA procedures are also implemented globally; the following characterization builds on the review of SEA in 12 selected countries.

Post Production Monitoring

Production monitoring is a specialized on-site inspection performed to verify the production status and the inventory of raw materials. An inspector examines our factory and checks the quantity of the raw materials and accessories required for your products before production.

Environmental Components of EIA air, noise, water and land

The components of environmental impact assessment include Water Environment, Biological Environment, Land Environment, Air Environment, Noise Environment, Socio-economic and Health Environment, EIA Risk Assessment and Environment Management Plan.

The Components of EIA (Environmental Impact Assessment)

EIA) evaluates the potential impacts of a project, plan, or development. The tool used to identify and assess a project's potential impacts on the natural and social environment and to recommend measures to mitigate those impacts. The EIA process in India involves several stages, including scoping, public consultation, preparation of an EIA report, review by an expert committee, and final decision-making by the MoEF. The components of environmental impact assessment include Water Environment, Biological Environment, Land Environment, and Air Environment, Noise Environment, Socio-economic and Health Environment, EIA Risk Assessment, and Environment Management Plan. The assessment considers a project's environmental, social, and economic impacts, including effects on air and water quality, biodiversity, soil, land use, cultural heritage, and human health. The results are used to inform decision-making and recommend measures to mitigate the project's negative impacts.

History of EIA in India

Environmental Impact Assessment (EIA) in India was first introduced in 1978 as element of the Water (Prevention and Control of Pollution) Act. The EIA process was initially voluntary, and it was only in 1994 that it became mandatory for specific projects to undergo EIA. In 1994, the Ministry of Environment and Forests (MoEF) issued the Environmental Impact Assessment Notification, which required all projects listed in the notification to undergo EIA before being granted environmental clearance. The notification was revised in 2006 to expand the list of projects that require EIA. The EIA process in India involves several stages, including scoping, public consultation, preparation of an EIA report, review by an expert committee, and final decision-making by the MoEF/ SEIAA.

Over the years, there has been criticism of the EIA process in India, with some stakeholders arguing that stringent not enough and that it has not prevented environmental damage caused by projects. However, the EIA process continues to be an integral tool for ensuring that development projects are designed and implemented in a way that minimizes their impact on the environment and communities.

Importance of EIA

Environmental Impact Assessment is vital in promoting sustainable development and protecting the environment. Here are some of the key reasons:

Identifying Potential Impacts: EIA helps identify the potential environmental, social, and economic implications of a proposed project before implementation, which enables decision-makers to make informed choices about whether to proceed with the project and to identify measures to mitigate any negative impacts.

Promoting Sustainable Development: EIA helps ensure that development projects are designed and implemented in an environmentally sustainable, socially responsible, and economically viable way.

Protecting the Environment: EIA helps identify and assess a project's potential environmental impacts and recommend measures to mitigate those impacts. This helps to protect the environment, including air, water, land, and biodiversity.

Enhancing Public Participation: EIA requires public consultation and participation, which helps stakeholders' concerns and perspectives are considered in the decision-making process.

Improving Decision-Making: EIA provides decision-makers with information about the potential impacts of a project, enabling them to make informed choices about whether to proceed with the project and to identify measures to mitigate any negative effects.

Components of an Environmental Impact Assessment (EIA)

The components include Water Environment, Biological Environment, Land Environment, Air Environment, Noise Environment, Socio-economic and Health Environment, EIA Risk Assessment, and Environment Management Plan.

The components of environmental impact assessment are as follows.

Sl. No.	Components
1	Air Environment: This component assesses the impacts of project on the air quality, including emissions from machinery, traffic, and other sources.
2	Noise Environment: This component assesses the project's potential impacts on the ambient noise levels in the surrounding area and evaluates measures to mitigate noise pollution.
3	Water Environment: This component assesses the impacts of the project on the water quality, including impacts on rivers, lakes, wetlands, and groundwater
4	Biological Environment: This component assesses the project's potential impacts on flora and fauna of the surrounding area and evaluates measures to protect biodiversity
5	Land Environment: This component assesses the project's potential impacts on the land use patterns and soil quality and evaluates measures to protect land resources
6	Socio-economic and Health Environment: This component assesses the impacts of the project on the social conditions of the surrounding area, including impacts on livelihoods, public health, and community well-being
7	EIA Risk Assessment: This component evaluates the potential risks and hazards associated with the project and recommends measures to manage and mitigate those risks.
8	Environment Management Plan: This component outlines a plan for managing and monitoring the project's environmental impacts during construction, operation, and eventual decommissioning. It also includes measures to ensure compliance with environmental regulations and address potential environmental issues that may arise during the project lifecycle.

Composition of Expert Committee

Member-Secretary-a serving officer of the respective State Government/ Union territory, who is familiar with environmental laws. Chairperson – an expert in EIA process with a term of 3 years. Non-officio Member - an expert in EIA process with a term of 3 years.

The Committees will be made up of experts from the following fields: Management of Ecosystems, Control of air/water pollution, Management of water resources, Conservation and management of flora/fauna, Planning for Land Use, Rehabilitation/Social Sciences, Project evaluation, Ecology, Subject Matter Experts in Environmental Health, Representatives of non-governmental organizations (NGOs) and individuals concerned about environmental issues.

Explanation

The Chairman will be a distinguished and experienced ecologist, environmentalist, or technical professional with extensive managerial experience in relevant development. The Impact Assessment Agency's representative will serve as Member-Secretary., Except for those specifically nominated as representatives, the chairman and members will serve in their individual capacities, A committee's membership cannot exceed 15 people.

EIA means Environmental Impact Assessment. It is a tool for anticipating environmental impacts of proposed development activities and recommending mitigation measures and strategies.

Steps in EIA Process

Stages	Description
1. Screening	Screening is done to find out which projects or developments require assessment study
2. Scoping	It finds out impact keeping harmony with legislative requirements, international conventions, expert knowledge and public involvement and also to identify alternative solutions
3. Assessment	Determines impacts and development of alternatives, to predict environmental hazards.
4. Report submission	Submitting EIA report, including an environmental management plan (EMP), and a non-technical summary for the general audience
5. Review similar cases	Mapping EIA study reports
6. Decision	Finalization of approved decision whether will have project or not. the project or not.
7. Monitoring	T examine expected impact and reality in field. The work is done or not according to approval.

Impact Prediction

Impacts are determined and taken for action. Impact prediction is designed to identify the degree of impacts, and provides the basis for the assessment of significance. Impact evaluation helps assess the relative significance of impacts.

Mitigation Measures in EIA

Mitigation measures are means to prevent, reduce or control adverse environmental effects of a project, and include restitution for any damage to the environment caused by those effects through replacement, restoration, compensation or any other means.

EIA Report: EIA report makes sure that project decision makers think about the likely effects on the environment at the earliest possible time and aim to avoid, reduce or offset those effects.

The EIA report is a statement of the likely impacts of a proposal and how these can be mitigated and managed. It is a decision document, not a compendium of technical information. As such, the EIA report should be both rigorous and easily understood.

What is EIA Format?

EIA template is a framework used to standardize the determining and evaluating the effects of a proposed project on the environment.

Screening, Scoping, Collection of baseline data, Impact Prediction, Mitigation Measures EIA Report

Screening

Screening is the deciding on whether an EIA is required. This may be determined by size (e.g. greater than a predetermined surface area of irrigated land that would be affected, more than a certain percentage or flow to be diverted or more than a certain capital expenditure).

ii) Scoping:

Scoping is a critical, early step in the preparation of an EIA. The scoping process identifies the issues that are be of most importance and eliminates those that are of little concern. Typically, this process concludes with the establishment of TO (Terms of Reference) for the preparation of an EIA.

Collection of Baseline Data

Baseline Study in EIA process usually includes data collection and analysis on various aspects of the project area, such as: Physical Environment (e.g., topography, geology, soil, water, air quality, and climate). Biological Environment (e.g., flora and fauna, habitat, biodiversity, and ecological processes).

Impact Prediction

For prediction the impacts are identified and taken for action. Impact prediction is designed to identify the intensity of impacts, and provides the basis for the assessment of significance. Impact evaluation is a process that helps assess the relative significance of impacts.

Mitigation Measures in EIA

Mitigation measures are means to prevent, reduce or control adverse environmental effects of a project, and include restitution for any damage to the environment caused by those effects through replacement, restoration, compensation or any other means.

EIA Report

EIA is a tool used to estimate the significant effects of a project on the environment. EIAs make sure that project decision makers think about the likely effects on the environment at the earliest possible time and aim to avoid, reduce or offset those effects.

The EIA report is a statement of the possible impacts and actions for mitigation. It is a decision document, not a compendium of technical information. As such, the EIA report should be both rigorous and easily understood.

What is EIA Format?

EIA template is a framework used to standardize the method of determining and evaluating the effects of a proposed project on the environment.

Public Hearing, Decision Making Monitoring and Implementation of Environmental Management Plan

Public Hearing

A Public Hearing during EIA will provide an opportunity for concerned stakeholders to express opinions, voice their concerns, and give suggestions to the authorities to encourage fair decision-making. The benefits and effects of a project are laid down for the public before the project starts.

The general public has a right to knowledge and to participate in discussions on issues that could impact people's lives, resources, and property. A Public Hearing during EIA provides a venue for interested parties, those who will be affected, and/or entire communities to learn about the project and exchange pertinent information.

History of Public Hearing

The EIA notification of 1994 was amended in 1997 to require a public hearing in the process of EIA for development projects. Under the Environmental (Protection) Act, environmental impact assessments (EIA) were introduced in India. However, it became legally binding when the Ministry of Environment and Forest **(MoEF)**[1] issued the EIA notification on January 27, 1994, in accordance with EPA 1986, making environmental clearance necessary for any activity's extension or modernisation for the establishment of new projects. Amendment to this notification in 2003 mandated that public hearing was not required for offshore exploration activities beyond 10 km from the nearest habitation, village boundary or ecologically sensitive areas such as mangroves (with 1,000 sq. mt area), coral reefs, national parks, sanctuaries, reserve forests, marine parks and breeding and spawning grounds of marine life.

Stages in the Public Hearing Process

Public Hearing during EIA comprises of two aspects; a public consultation process in which only the local -affected people can participate and a process for obtaining written comments from those who are concerned citizens.

Stage	Activities
Stage 1	The consultation process has two components. First, it is carried out at the site or in its proximity- district-wise, in the manner prescribed in Appendix IV of the notification for ascertaining concerns of local affected persons. Secondly, it aims to obtain responses in writing from other persons having a plausible stake in the environmental aspects of the project. Anyone affected by the project is entitled to have access to the Executive Summary of the EIA. The following people are allowed to make written/ oral suggestions in public hearing. 1. Local residents 2. Local associations 3. Environmental groups: active in the area 4. Any other person located at the project site/s of displacement
Stage 2	The SPCB forms a Public Hearing panel with a representative from the SPCB, the district collector or their nominee, representatives of the state government handling the project, a maximum of three local Panchayat/municipal representatives, and a maximum of three senior citizens nominated by the district collector.
Stage 3	The project proponent will be required to submit 20 sets of documents as prescribed in the 2006 notification at the Sub-Regional office/Regional office of the concerned SPCB. This will include the executive summary of the project in English and the local language, including information about the likely environmental problems and proposed action for controlling degradation and pollution from those problems on the environment.
Stage 4	The project developer will get a NOC from the SPCB following the conclusion of the Public Hearing during EIA and submit an application to the secretary of the MoEF for environmental clearance.

Monitoring and Implementations of Environmental plan

Monitoring is a planned and systematic collection of environmental data to meet specific objectives and environmental needs. To ensure that mitigation measures recommended in the EIA are implemented and maintained throughout the operational life of the project.

Monitoring provides data on the environmental and social impacts of the project for the whole project lifecycle. As part of their operations, most development projects involve regular monitoring of indicators (including such things as quantity of ores extracted, materials processed, energy used and sewage released, etc.). More specifically, the information collected during monitoring activities helps to ensure that the priorities listed in the Environmental Management Plan (EMP), mitigation measures, and contingency plans are properly implemented, and that these plans and measures are effective in addressing the project's impacts.

Monitoring is critical to ensuring the fulfilment of all the commitments made in the approved EIA. It is one of the ongoing outcomes of EIA for a given project. Monitoring is also important for any changes that happen and in communities by the project and other local and/or global events, such as changes in livelihoods due to economic crisis or migration, or differences in water availability due to drought. After the project is implemented, basic monitoring efforts will continue during project remediation.

Monitoring is usually carried out by the project proponent, under the supervision of independent agencies and/or government agencies. The key steps in creating a monitoring plan are:

i) Identify the focus areas to be considered for the plan (based on the impacts, mitigation measures and other issues from the contingency plans).

ii) Select of a long list of indicators for the focus areas, such as those on population, health, and natural elements used by the population (soil, drinking water, etc.).

iii) Prioritize a core set of indicators based upon agreed criteria.

iv) Identify data collection protocols.

v) Develop data interpretation methods and create a reporting template.

vi) Schedule activities, evaluate program and personnel costs, define personnel responsibilities.

Implementations of Environmental Plan

A good environmental management plan should be a living document. Initially, the plan needs to be designed to achieve your goals based on the information available. As you implement the plan and document the results, pay attention to how tasks are executed and how effective they are.

To Implement a Successful Environmental Management plan, consider three things.

Know What is in our plan

Presumably, we know what commitments have been made in our environmental management plan. But do er know what other commitments have been made in other documents associated with our plan? Are they consistent with what is in our plan? Too often multiple, sometimes conflicting, commitments are made by different people in different documents during the rush to permit or develop a project. If our environmental management plan has not resolved these conflicts before implementation, it may not be possible to do so. Developing a register of environmental obligations that documents all of the commitments made for the project early in the planning stages can reduce the potential for overlaps and conflicts. It can also help identify gaps that may exist in our management measures.

Make Sure Your Plan is Practical

Environmental plans look great on paper. We have identified key issues, defined actions to mitigate environmental risks, laid out a clear schedule and budget and it appears that all the angles are covered. But it is important to ask, "Is this plan something that we can actually implement on the ground?"

Our plan has to be sensible and realistic for the specific circumstances. We need to consider what technical, economic, and human resources will be needed to execute the plan.

Review and Revise Plan Constantly by Using Evaluation Tools

A good environmental management plan should be a living document. Initially, the plan needs to be designed to achieve our goals based on the information available. As we implement the plan and document the results, pay attention to how tasks are executed and how effective they are. This information can then be used to determine how effective the plan is at achieving the goals. Then we can adapt the plan as needed. A strategic plan is to be able to evaluate our success with measurable metrics. Make what works our standard practice and adjust the elements that aren't working well. Write and implement environmental plan with continual improvement in mind.

Assessment of Alternatives Delineation Mitigation Measures and EIA

Alternatives assessment is a process for identifying, comparing, and selecting safer alternatives to chemicals of concern (including those in materials, processes, or technologies) on the basis of their hazards, performance, and economic viability [Massachusetts Toxics Use Reduction Institute (MA TURI) 2013].

According to Simonson and others, there are three approaches in alternative assessment: Authentic assessment, performance-based assessment and constructivist assessment.

Authentic Assessment

Authentic assessment is the idea of using creative learning experiences to test students' skills and knowledge in realistic situations. Authentic assessment measures students' success in a way that's relevant to the skills required of them once they've finished your course or degree program.

We have distilled this theory down into the four pillars of great assessment: purpose, validity, reliability and value. The Four Pillars of Assessment resource guide will provide us with a strong understanding of what underpins each pillar and how it supports great assessment.

Performance-Based Assessment

Performance-based assessments are an objective way to monitor learners' levels of understanding. Find out everything we need to know about this approach to decide which type of assessment fits your needs.

Performance-based assessment requires students to demonstrate or apply their knowledge, skills, and strategies by creating a response or product or doing a task.

Constructivist Assessment

Constructivist approach to assessment is a formative rather than summative. Assessment and Evaluation in constructivism focuses on the process that the individual learner takes in the process of knowledge construction rather than just the product.

Delineation of Mitigation Measures

Mitigation measures proposed herein are the result of an iterative process that took place between the environmental effects prediction and the engineering design teams.

Based on preliminary environmental effects results, further mitigation measures were incorporated into the design of the Project to ensure the protection of the physical, biological and human environments.

Objectives The proposed mitigation measures provide the basis for the development of environmental management plans and monitoring programs for the Project, required to meet Federal and Provincial environmental approval and permitting requirements and conditions as well as IAMGOLD policies of environmental compliance.

The mitigation measures aim to:

1. Protect the physical, biological and human environments;
2. Manage mineral wastes; manage hazardous compounds and wastes; and
3. Provide the basis for the development of monitoring plans.

Mitigation measures are means to prevent, reduce or control adverse environmental effects of a project, and include restitution for any damage to the environment caused by those effects through replacement, restoration, compensation or any other means.

Based on environmental management plans and monitoring will allow for continual assessment of the effectiveness of these mitigation measures. As new information becomes available through these monitoring programs, selected mitigation measures may be revised if they prove less effective than anticipated.

Notification Environmental Clearance, Rejection, Participation of EIA

Notification Environmental Clearance

The Environment Impact Assessment (EIA) Notification, 2006, is the governing legal instrument to grant green clearance for the establishment or expansion of an industry on the basis of the potential environmental impact of the project.

On which notification EIA is made mandatory in India?

On 27 January 1994, the Union Ministry of Environment and Forests (MEF), Government of India, under the Environmental (Protection) Act 1986, promulgated an EIA notification making Environmental Clearance (EC) mandatory for expansion or modernization of any activity or for setting up new projects listed in Schedule.

The EIA Notification of 1994 made the clearance mandatory for all new projects and expansion/modernization of existing projects covering 29 disciplines (later increased to 32) which included hydro-power, major irrigation and flood control projects.

Rejection

The most common causes of rejection include submission of application packages with incomplete information, e.g., subject not including the company submitting the investigation request as a current employer, missing SSN for spouse or co-habitant, fingerprint cards, information for relatives and failing to provide.

Participants of EIA

Participants

1. EIA applies to private and public sections.
2. Those who propose the project.
3. The environmental consultant who prepares EIA on behalf of project proponent.
4. Pollution Control Board (State or National).
5. Public has the right to express their opinion.
6. The Impact Assessment Agency.

Shortcomings of EIA and how to improve EIA process

Shortcomings of EIA Process

Loopholes in the Application Process

Loopholes in the application process are one of the shortcomings of Environmental Impact Assessment because there are many development projects which have a severe impact on the environment. Still, they were relieved from the notification process. This happens when they are not listed in Annex I or their investments are not adequate to send notice according to the current regulations.

Inadequate Public Participation

There is little scope for public participation and accounting for public opinion in the EIA process, and indigenous knowledge and understanding often get neglected in the evaluation process. Therefore, there are numbers of instances where these limitations create much bigger obstacles to the progression of the projects.

Insufficient Expertise

The lack of expertise in different fields to evaluate the impact is one of the shortcomings of EIA. Often there is an inadequate standard of control or lack of expertise on the part of environmentalists, bio-diversity experts, social scientists, and social workers. That creates a lack of understanding of the evaluation process of EIA.

Dependency on Technical and Scientific Data and Analysis

The EIA process relies heavily on technical and scientific field data. Technical data, data collection processes, and measures are not always accurate or can paint a true picture. Macro data without proper in-depth study often lacks nuance, leading to errors and miscalculations in evaluation.

Quality of EIA

Normally the EIA process is financed by individual investors or organisations and agencies. Agencies without strong oversight might want clearance for the project and thus ignore unfavourable aspects. So, EIA reports often try to overlook or misjudge the reality of the environment. Further, it lacks many important details from the ground. Many EIA reports are defective and partial, and many reports only submit one seasonal report in the process. These shortcomings in EIA often raise questions about the quality of EIA.

Scope to Improve the EIA Process

Independent Body of EIA

There is long-standing demand from civil society and environmentalists that EIA should have an independent functional entity. The EIA conducting body should be created by experts from the judiciary, environment science, social science, anthropology, people from communities and local bodies. These EIA structures will function independently, not abiding with the Ministry of Environment and Forest.

Sector wide EIA

An action must be taken to set up new policies for sector wide EIA to understand the impact and repercussions of not just a single project but different projects and how they impact each other. This requires carefully deliberated parameters and critical observation and analysis.

Strong Policies

To further strengthen the EIA, policies should be implemented that would enforce every developmental project to go through EIA scrutiny without any deviation. It is also vital to implement strong policies to stop industrial or other development projects in environment-sensitive areas.

Public Participation

EIA should be more inclusive towards public hearings and participation to consider public knowledge to resolve social and environmental consequences. If any development happens in the forest or marine area, to save natural flora and fauna.

References

Abelson, Robert, P. (1968) “Simulation of Social in Behaviour” in G. lindezyand E. Aronon (eds) Hand book of Social Psychology, Vol. II, Mass: Addition-Wesley, 1968, p. 275.

Adler, P.A. and Adler, P. (1994) Observation Techniques, Sagar Publication.

Bernard. (1952) Content Analysis in Communication Research, Glencare, III. The Free Press.

Black, J. ames, Champion, DEAN, J. (1976) Methods and Issues in Social Research ; Authors, James A. Black, Dean J. Champion ; Edition, illustrated; Publisher, Wiley.

Clover, M.R. and Nagel, E. (1934) An Introduction to Logic and Scientific Method, New York.

Creswell, John W. (2002) Research Design Qualitative, Quantitative and Mixed Methods and approaches, Thousand Oaks, CA: Sage Publication.

Emery, C. Willium (1976) Business Research Methods. Homewood, Illinois: Richard, D. Trwin.

Emory, C.W. & Cooper, D.R. (1991) Business Research Methods (4th edn. Boston, M.A. Irvin).

Encyclopaedia of Social Sciences (1973) Vol. IX. The Mac Million Company.

Fred, N. Kerlinger (1978) Foundation of Behavioural Research. Surjeet Publication, Delhi 7.

Gouri K.Bhattacharya and Ricjard A. Johnson (1977) Statistical Concepts and Methods. John Wiley & sons.

Krishnaswamy, O.R and Rangnathan, M (2005) Methodology of Research in Social Sciences, Himalaya Publishing House, Delhi.

Lastrucci, C.L. (1963) The Scientific Approach- Basic Principles of Scientific Method, Cambridge, Mass, Schenkman.

Levein, Irwin, P. (19999) Relating Statistics and Experimental Design. Thousand Oaks CA: Sage Publication.

Lundberg, G.A. (1942) The operational Definition of Social Science. American Journal of Sociology pp.727-745.

Miller, F. (1986) Use, Appraisal and Research; A case study of Social History. The American Archivist 49(4), 371-392.

Myrdal, Gunnar (1970) The Changing World Poverty. Hammond’s Worth, Penguins book (eds).

Nargundkar, R. (2007) Marketing Research: Text and Cases, Tata McGraw Hill New Delhi.

Orenstein, Alan & Phillip,William RF (1978) Understanding Social Research-An Introduction, London Eisenhardt, K.M. (1989) Building Theories from Case Study Research .Academy of Management Review 14(4), 352-550.

Pavako, Ronald, M. (1967) Sociology of Occupation and Profession, Itasci, F.E. Pacock.

Ray, G.L. and Sagar Mandal (1999) Research Methods in Social Sciences and Extension Education, Naya Prakash, Calcutta.

Ross, J and G.Alloport (1967) Personal Religious Orientation and Prejudice, Journal of Personality and Social Psychology, 5(4), 432-443.

Seltz C. (1960) Research Methods in Social Relation, USA, Methuen & Co. Lted.

Shadish, W.R., Thomas D. Cook and Donald T. Campbell, D.T. (2002) Experimental and Quasi experimental designs for generalized causal inference. Boston: Houghton-Miffilin. An update of a classic by a third author.

Thomas, W.L. (1909) Sources of book for Social Orgin, Chicago University of Chicago, press.
Wolf, D. (1940) Factor Analysis 19400 Factor Analysis, Chicago University, Chicago Press.
Young, Paulin, V. (1977) Scientific Social Survey and Research, New Delhi, Printice–Hall, India.

Index